2010 欧博设计

主编寄语
EDITORIAL

《梦想-现实》

梦想

有关梦想的描述古今中外比比皆是，最知名的也许是马丁·路德·金的《我有一个梦想》。

一个有梦想的人犹如火车有了轨道，会为了自己的目标与追求，不畏坎坷，一路跋山涉水去收获痛快淋漓的无悔人生。

——有梦想的人，人生是美好的

梦想并非是空想和不切实际的狂想。梦想是一种积极的生活态度，是对一种适合自己但在当下并非能轻易变为事实的生活状态的追求。梦想可以营造行为与灵魂对话的精神气场，可以唤醒沉睡不知的潜能，唤醒对生活的热爱和对未来的憧憬。如果有梦想，我们就不会过于在意世俗的纷扰；如果有梦想，我们的生命就会充满旺盛的动力；如果有梦想，我们的生活就会充满无限的希望与可能。

——有梦想的人，人生是顺畅的

生命充满甜蜜、喜悦、失意、迷茫，是什么，才能让我们安然度过这悲喜交加、起起落落的人生？我想只有源自内心的召唤，只有梦想的力量才能支撑起我们这繁杂世界。梦想的力量是神奇的，它会引领你改变自己，完善自己。如果你拥有一个梦想，也许你会发现敏感纤细的内心正一天天变得更加强大。

——有梦想的人，人生是快乐的

因为有梦想，你能拨开黑夜的迷雾看到未来的曙光，能在绵绵阴雨中感受春天的气息，能把艰辛的设计当作更新自我、挑战自我、完善自我的历程。你会感到快乐并感染身边的人，当每个人的身心都被快乐所充盈，一切问题都会迎刃而解。所有的这一切改变，只因你拥有一个梦想！

现实

哲人说"存在是合理的"，我更愿意理解为，人存于世，永远无法回避周围的人与事；一味的逃离，将无法感受这个鲜活世界的脉动，也不会感到真正的快乐。

——理解现实的人，生活是充实的

每个人的生活都是具体而平凡的，高潮有时、低潮有时、成功有之、失败有之……对于理解现实的人来说，不管是什么样的人生境遇都是一种体验与历练。人降临于世是偶然，而人生是他必须面对的必然。他不会抱怨，用真心感恩；决不妥协，去努力改变；不甘于平凡，勇于创造适合自己的美好人生。

——敢于深入现实的人，生活是成功的

当你准备跨入现实，可能会感到惶恐、困惑、无助，但只要敢于步入现实，现实就是你真正的导师，让你发现未知的自我。如果你足够勇敢，现实会给你切实的回报——顺利的生活、成功的事业！

——关切于现实的人，与世界是和谐的

既然无法脱离现实，那我们就直面现实，关切现实中的一切人与事，建立与现实相连的纽带，并在编织这条纽带时，考究其用料，思量其方法，适度用其力，那么你与现实、现实与你一定圆融无碍。

现实与梦想是两个场域，不可分离。无视现实的梦想犹如空中楼阁，没有梦想的现实犹如一潭死水！
相信梦想终会实现，现实终会美好！
最后，我引用星云大师的几句话与大家共勉：
　忙时井然，闲时自然；
　顺多偶然，逆多必然；
　得之淡然，失之坦然；
　褒则常然，贬则泰然；
　悟通八然，此生悠然。
感谢大家一年来的辛劳与付出，感谢所有一直予以我们信任与支持的你们！祝大家新年快乐，并祝福大家在新的一年里美梦成真！

欧博设计
董事长　主持设计师　冯越强

DREAM & REALITY

The Dream
Dream exists anywhere. But the most astonishing is Martin Luther King's I have a dream. Those with dreams like the running train on the right track. With goal, we will not hang back. We pursue a colorful, unconstrained and regretless life.
Passionate Dream, Wonderful Life
Dream is not intangible imagination. It is a positive life attitude. It ascertains suitable condition when we feel tough and unstable. Dream cultivates a spiritual context where action and soul can have a conversation, where potential is evoked and where passion for better future is triggered out. Holding a dream while neglecting those mundane revitalizes amazing talent. Through whole life, even a small and ordinary dream can make life wonderful, infinite and hopeful.
Inspiring Dream, Refreshing Life
We experience success, sweetness and delight. But, we also suffer from frustration, confusion, pain and unease. What can help us get through those making life sad and pessimistic? The answer is dream. If you have dreams, you might discover a way proceeding to tolerance, happiness and excellence. The power of dream is miraculous. It leads you to change and perfect yourself no matter which environment you are in.

Colorful Dream, Diverse Life

The dream makes you look through the night fog to see tomorrow's sunshine and accumulate spring flavor in gentle rain; the dream makes you regard the hard creation process as self-renewal, self-transcendence and self-perfection. Therefore, you will feel happy and your happiness will moved people around you. When we are filled with joy, all the problems can be easily solved and you may find everyone amiable. All in all, the change takes place just because you have a dream.

The Reality

A wise man once said, "Everything is right in the world!" The way I make sense is that we cannot be isolated from the surrounding. Those trapped in a cell will neither enjoy the unique scenery in the world nor the true joy of life.

Make sense of the world, fertilize the life. Life is a symphony with ups and downs rhythm resembling a vivid record of maturation. If a person tries his best appreciate the gentle tune, he may reach the funny and refreshing layers of life, equip each day with beauty, cultivate the mind with dynamics, and fulfill heart with joy. For them, life is a no returning road. When encountering difficulties, they do not hesitate or complain. Rather, they make all hurdles as part of a life with a thanksgiving mind. We are ordinary people. But our nature should not be mediocre or even deteriorated. Dream is human courage in a man: to bear unflinchingly what heaven sends. I believe that all our dreams can come true if we have courage to pursue them.

Facing reality, leading to success

Risk is unavoidable. If you want to achieve something, we have to advance bravely. Even if suffering from difficulties or frustrations, we should take them in stride. Enjoy a tough but energetic day; To Be strong enough to accept the challenges of life. Even if we take small steps during the difficult periods, finally you will advance to destination. Connected with people, harmonious with the world.

We can't escape the cruelty of social facts; we must face it with braveness. Being considerate with people and everything around, we will build meaningful ties with the world. If you weave the ties attentively, your relationship with the reality will become as harmonious as fiddle-strings. Dream and reality are two realms but connected together. The dream divorced from the reality is merely a castle in the air; the reality devoid of the dream is like a pool of stagnant water. Believe that the dream will come true, the reality will be magnificent.

Finally let's share the wisdom of by Venerable Master Hsing Yun by his poem:

Work with order, live in leisure.

Moved by second, trapped as constant.

Get as light, loss no sigh.

Praised then ceased, hit but eased.

Grasp the core, life no sore.

Appreciate the contribution and support that everybody makes for the company! Wish All a Happy New Year!

May all your dreams come true in the new year!

AUBE Conception
President & Chief Designer: Yves Yueqiang Feng

前言
PREFACE

这是一本2010年的主要工作备忘，是21世纪第一个十年的末端，欧博全体奉献给您的一份礼物。礼物的送出者是书中的这群人。如同他们各自的面容，差异性显而易见，但富有理想、勇于实践、持之以恒又是这群人的共性。

我们主张对生存境遇的现象学解读，存在本身是边缘性、生发性偶在。存在先于本质，我们生产自身。目的与过程相较，后者更有意义。跨界是我们现有的生存状态。在这样一个阻遏自我身份的时代，"身份"是什么？我们愿意共同寻求。

设计如同哲学，挣脱不出语言的牢笼，语言既不能证明形而上学本体存在，也不能证明形而上学本体不存在，语言的"两不性"揭示了哲学与设计日趋表层化的缘由。我们运用工具，工具占有我们，维特根斯坦说"世界被把握成图像了"，万维网、图像化、数字化已经成为我们的存在方式，我们就像放在镜箱中的变色龙，究竟该作如何反应？董豫赣讲"文学将要杀死建筑"，建筑还有本体可以追寻吗？我们愿意相信有。

现今，建筑师与时装设计师之间可拿等号连接，但脱掉霓裳的裸体建筑更能让我们触及他（她）的质感和温度。卡尔维诺曾留下《千年文学备忘录》作为文学箴言警醒后辈，我们作为自身生活的设计者到底应否一味盲目地沿着技术大道一路狂飙？我们愿意说不。

在第二个十年的门前，适逢姜文电影《让子弹飞》轰然上映，"让子弹飞一会儿"是电影中特有魅力的一句台词，飞翔中虽目标仍待锁定，但疾飞的过程就是子弹的宿命。"机器生物化，生物工程化"，这是凯文-凯利（Kevin Kelly）在20年前的名作《失控》中对世界的预言，值得快慰的是，我们工作与生活一直践行的恰是"蜂群思维"和"众包模式"。

新的千年刚刚开始，欧博保持并形成一个更大的裹挟开放的子弹蜂群，是所望焉。谨序。

欧博设计

设计董事　白宇西

This yearbook is contributed by the entire AUBE staff to review the group's achievements in 2010, the end of the first decade of the 21st century. Behind the achievements are diversified faces on the title page, who have gathered at the AUBE for common aspiration, courage and persistence.

We have been dedicated to a phenomenological interpretation of human's survival circumstances, since existence is in essence a marginal and originating accident. Existence precedes essence and we create our own values, determining a meaning to our life. Therefore, process is more meaningful than purpose. Having a crossover mind, we have been willing to seek "identity" of architecture in the current times self-identity is either ignored or covered.

Like philosophy, design is confined to a cage of language, which can not prove or disprove existence of ontology. Language's such inabilities cast light on a superficial trend in philosophy and design. We employ tools but end up in being occupied by tools. The renowned Austrian philosopher in the twentieth century, Ludwig Wittgenstein, said that the world had been interpreted as images. Like a chameleon in a mirrored box, what should we do in response to a way of being, which features Internet, visualization and digitalization? "Literature is killing architecture," said Dong Yugan, a well-known Chinese architect. However, we are still wiling to believe in existence of ontology of architecture.

Nowadays, architects can be likened as fashion designers. However, nude architectures not in extravagant dresses make it easier for us to feel their texture and temperature. Italian writer, Italo Calvino, used to caution younger generations in "The Six Memos of the Next Millennium" that, as designers of our own lives, should we gallop blindly on a technology road? We are willing say "No."

"Let bullets fly a while" is a charming lines in the newly-screened " Let Bullets Fly", a film by talented Chinese director, Jiang Wen. A flying bullet still needs to lock its target, but the scudding process is the destiny of the bullet. "Machines turn biological and living beings turn engineered", foretold by the American writer, Kevin Kelly, in his well-known "Out of Control" 20 years ago. Gratifyingly, we have been employing "hive mind" and "crowdsourcing mode" in our work and lives.
The AUBE will build up its cohesive and open-minded bullet hive at the beginning of a new millennium.

AUBE Conception
Design Director : Bai Yuxi

目录
CONTENTS

URBAN 规划

012 贵阳云岩区渔安安井片区城市设计
Planning and Architectural Design of Yu'an & An Jing of Yunyan District in Guiyang

018 武汉新区四新生态新城"方岛"区域城市设计
Urban Design of "square Island" of Sixin Ecological Town in the New Area of Wuhan

022 台湾高雄海洋文化及流行音乐中心新建工程国际竞赛
Architectural design of Kaohsiung Maritime Culture & Popular Music Center

028 佛山西站综合交通枢纽概念性规划建筑设计
Planning and Architecture Design of Foshan West Station

034 长春南部新城净月西区生态商务金融中心（EBD）城市设计
Urban Design of EBD of Jingyue District, Changchun

036 花溪山水度假旅游城贵阳花溪概念性城市设计
Conceptual Urban Design of Huaxi Tourism Resort of Guiyang

040 中粮地产（集团）深圳宝安61区概念性规划建筑设计
Conceptual Planning and Architectural Design of Zone 61 of Cofco Real Estate, Bao'an District, Shenzhen

044 贵阳星云家电城项目城市设计及单体建筑概念性方案设计
Conceptual urban Planning & Architectural Design of Xingyun Home Appliances City of Guiyang

050 航天成都城上城规划、建筑设计
Planning and Architectural Design of Aerospace Town in Chengdu

056 深圳半岛城邦三四五期详细蓝图修编
Detailed Blueprint Modification of Peninsula Residential Community in Shenzhen

060 福州中央商务中心安置房规划建筑设计
Planning and Architecture Design of Resettlement Housing of Fuzhuo CBD

ARCHITECTURAL 建筑

066 珠海港珠澳大桥香港口岸国际概念设计竞赛
HongKong-Zhuhai-Macao Bridge HongKong Boundary Crossing Facilities International Design Ideas Competition

076 深圳市南山文化（美术）馆方案设计
Architectural Design of Nanshan Culture(arts) Museum of Shenzhen

082 深圳创维"半导体设计中心"建筑方案设计
Architactural Design of Skyworth Semiconductor Design Center of Shenzhen

090 深圳市高新技术企业联合总部大厦方案设计
Architectural Design of United Headquarter of High-Tech Enterprises of Shenzhen

094 深圳地铁一号线深大站综合体上盖物业方案设计
Architectural Design of Building Complex above Shenzhen University Station of Metro Line No.1

| 096 | 深圳移动生产调度中心大厦概念性方案设计
Architectural Design of Shenzhen Mobile Tower

| 102 | 深圳罗湖档案管理中心建筑设计
Architectural Design of Luohu Archives Center, Shenzhen

| 108 | 深圳实验学校中学部扩建工程方案设计
Architectural Design of Expansion Project of Shenzhen Experimental Middle School

| 116 | 深圳创业投资大厦建筑设计
Architectural Design of Vc & PE Tower of Shenzhen

| 118 | 深圳市青少年活动中心改扩建方案设计
Architectural Design of Shenzhen Adolescent Activities Center

| 124 | 深圳市鼎和国际大厦方案设计
Architectural Design of Shenzhen Dinghe International Tower

| 128 | 深圳南山建工村保障性住房建设规划建筑设计
Architectural Design of Resettlement Housing of Jiangong Village, Nanshan District, Shenzhen

| 132 | 深圳南海中学建设方案设计
Architectural Design of Nanhai Middle School of Shenzhen

| 136 | 河源深业东江波尔多皇家庄园规划建筑设计
Conceptual Planning and Architectural Design of Shum Yip • Bordeaux Royal Manor (Living Service Area of Dongyuan County of Heyuan, Guangdong)

LANDSCAPE 景观

| 146 | 贵阳云岩区十里花川公共空间及南明河两侧景观概念设计
Landscape Design of 10 Miles Flower Valley of Yu'an-Anjing District, Yunyan, Guiyang

| 152 | 昆明滇越铁路主题公园景观方案设计
Landscape Design of Yuannan-vietnam Railway Park of Kunming

| 158 | 贵阳国际会议展览中心西侧绿地公园景观设计
Landscape Design of Green Land Park on the West of Guiyang International Conference &Exhibition Center

| 162 | 深圳康佳研发大厦景观设计
Landscape Design of Konka R&d Tower of Shenzhen

| 166 | 深圳长富金茂大厦景观设计
Landscape Design of Finance Tower of Changfu of Shenzhen

| 170 | 深圳深业新岸线入口及商业街景观改造设计
Landscape Design of the Entry and Retail Street of Shum-yip New Shoreline Community of Shenzhen

| 174 | 贵阳中天世纪新城6、7组团景观设计
Landscape Design of Blocks 6&7# of Dream Land of Guiyang

| 180 | 贵阳中天世纪新城4组团幼儿园景观设计
Landscape Design for the Kindergarden of Block 4 of Dream Land of Guiyang

2010
URBAN
规 划

01 PLANNING
And Architectural Design of Yu'an & An Jing of Yunyan District In Guiyang

In our proposal, we try to exploit the tourism potentialities of Guizhou by taking full advantage of the natural resource such as thermal spring and to reinforce the sustainable development of the city to satisfy different demands of people and to create a happy city with different tourism themes and characteristics.

The main characteristic of urban design is the spatial form of "one axe, four lines and four gates".

One axe:
Along the Nanming River, main landscape axe, create a ten miles flower valley with local cultural characteristics and with biodiversity.

Four lines:
Natural culture line of Shuidong Road
Urban entertainment line of Beijingdong Road
Exhibition line of Dongerhuan Road
Living fashion line of Neihuan Road

Four gates:
Four skyscraper groups stands as four gates, respectively in characteristics of 'mountain', 'water', 'forest' and 'city'.

贵阳云岩区渔安安井片区城市设计

定位：
温泉，可充分发掘的基地资源，将提高资源含金量。
旅游，可参与联系互动，加强城区发展的可持续性。
城，满足多层次的需求，体现场市、街区、街道和谐氛围。

城市设计空间形态：
以"一轴、四线、四门户"为主要特征。
一轴——南明河景观的主轴，打造具有贵州文化、地域、生物多样性的强烈地表特征——"十里花川"
四线——水东路自然文化线
　　　　北京东路都市休闲线
　　　　东二环风貌展示线
　　　　内环路生活风尚线
四门户——以四组超高层建筑形成"山"、"水"、"林"、"城"四大门户

目标：通过本项目使贵阳成为贵州旅游产业的集散地；成为吸引西南地区、全国乃至世界，独特、高度集约的旅游之城；成为集旅游休闲产业各类主题和特征的城市"快乐源泉"。

Client : Guiyang Urban Planning Bureau
People's Government of Yunyan District of Guiyang
Zhongtian Urban Development Group Co., Ltd.
Location : Yunyan District, Guizhou
Land area : 755ha
Building area : 6 500 000m²
Function : Thermal spring tourism resort
Cooperators: Forrec Limited
　　　　　　Guiyang Urban Planning & Design Institute
　　　　　　Southwest Municipal Engineering Design & Research Institute of China
　　　　　　Wu & Song Associates (Shanghai), Ltd.
　　　　　　Guizhou Investigation Design & Research Institute of Water Conservancy & Hydropower
　　　　　　Guiyang Architectural Design & Surveying Prospecting Co., Ltd.

客　　户：贵阳市城乡规划局
　　　　　贵阳市云岩区人民政府
　　　　　中天城投集团股份有限公司
位　　置：贵阳市云岩区
用地面积：755hm²
建筑面积：650万m²
主要功能：温泉旅游城
合作单位：加拿大FORREC建筑设计事务所
　　　　　贵阳市城市规划设计研究院
　　　　　中国市政工程西南设计研究总院
　　　　　吴宋美加设计咨询上海有限公司
　　　　　贵州省水利水电勘测设计研究院
　　　　　贵阳建筑勘察设计有限公司

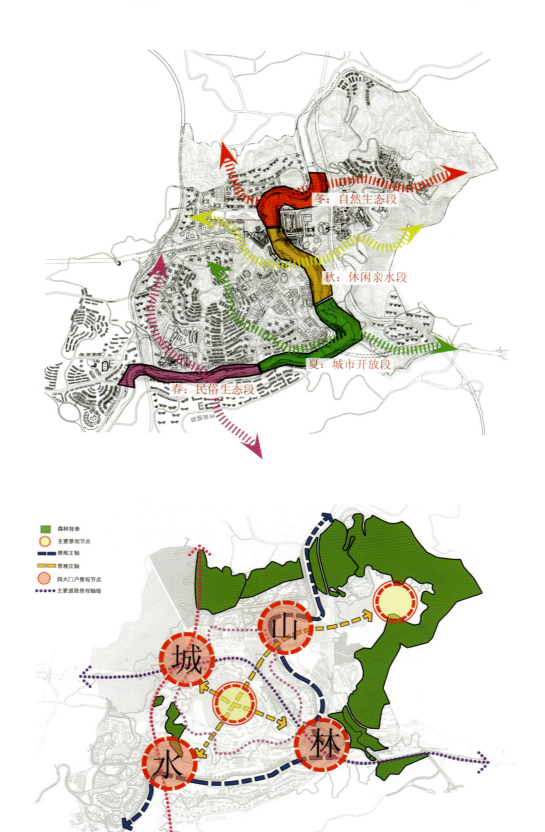

015 · PLANNING

EARTH ZONE
大地区

AIR ZONE
空气区

WATER ZONE
水区

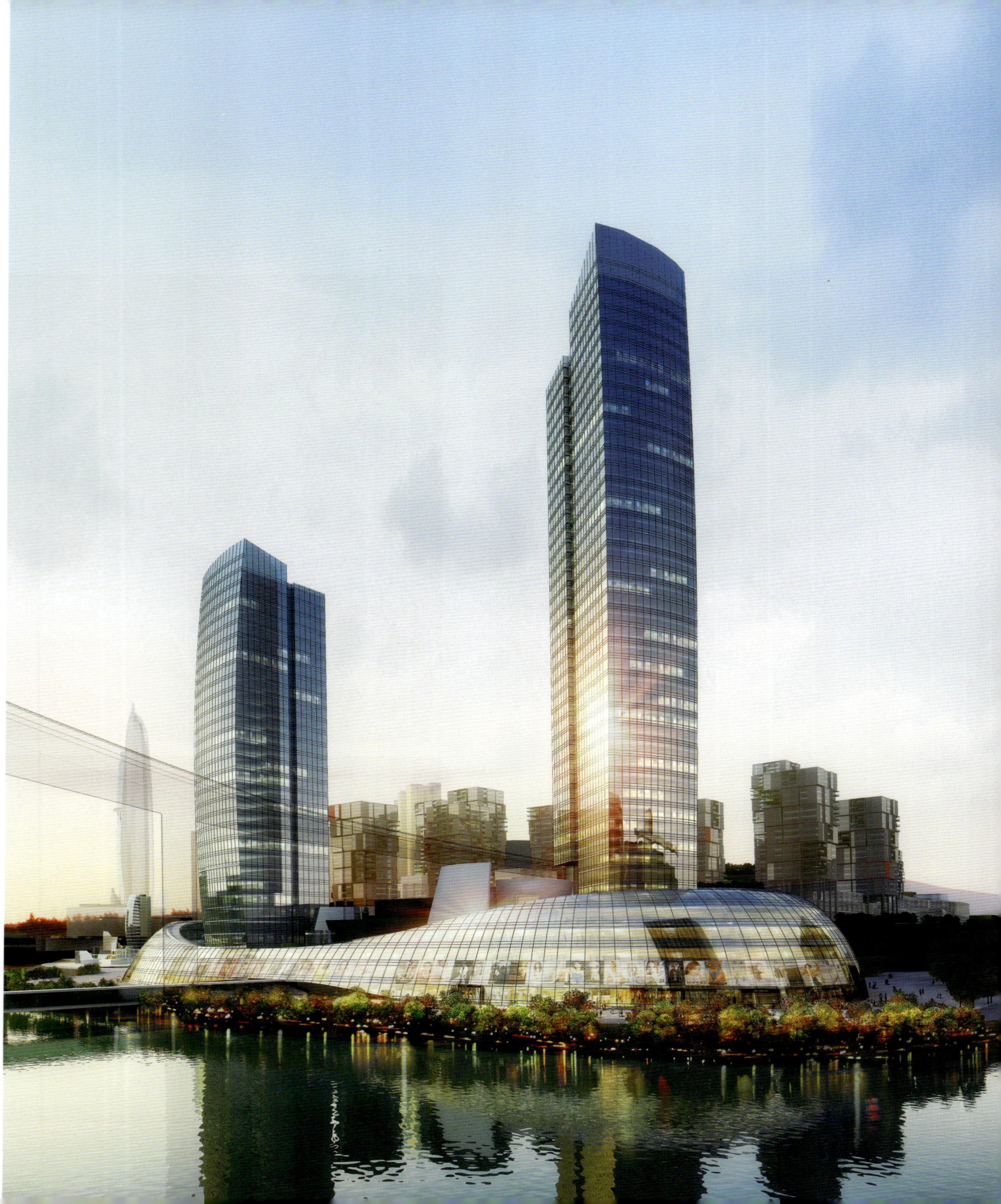

(FORREC)

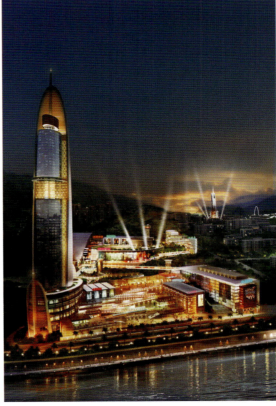
(FORREC)

02 URBAN

Design of "Square Island" of Sixin Ecological Town in the New Area of Wuhan

"Square Island" is located at Sixin district of Wuhan which is devoted for the development of functions such as exposition, conference, head office, finance, commerce, formation and specific service etc. It will be the producer service center of Wuhan and even of Central China and be new ecological town for residential community with high quality and waterfront landscape. The urban design aims for creating a brand new and ecological living community which could represent the landscape features of waterfront of Wuhan.

According to the principles and target hereinbefore, the functional distribution is divided as: 'seven cycles, two banks, four doors, one ring and two lines'. Because of this distribution, the 'Square Island' area looks like ripples in the city.

武汉新区四新生态新城"方岛"区域城市设计

方岛区域城市设计所在的武汉四新片区，依托武汉国际博览中心着重培育展览、会议、总部、金融、商贸、培训、专业服务等生产性服务新功能，致力打造服务武汉乃至整个华中地区的生产性服务中心，发展并形成市级副中心；同时建设高质量的居住区，形成代表武汉滨水景观特色的生态居住新城。

城市设计的目标定位为：在优势联系下，全新的、能够代表武汉滨水景观特色的、24小时生态复合大社区；建设以后工业社会为背景，以滨水文化为主题的优质生活区。

鉴于上述原则与目标，本区域的功能构成与分布概括为："七圈、两岸、四门户与一环两线"。顺应七个功能圈层，富于节奏感的城市空间开始呈现，整体区域将形成近似于"涟漪"的城市空间形态。

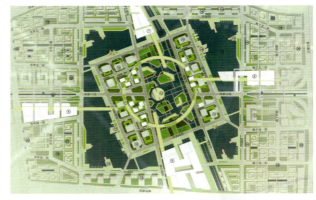

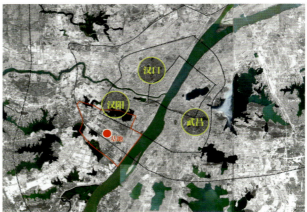

Client : Wuhan New Area Construction and Development Investment Co., Ltd.
Location : Sixin Ecological Town, Wuhan
Land area : 268ha
FAR : 1.5
Building area : 4 000 000m²
Height : <150m
Function : Complex of office, residence and commerce
International bidding : 2nd place

客　　户：武汉新区建设开发投资有限公司
位　　置：武汉四新生态新区
用地面积：268hm²
容 积 率：1.5
建筑面积：400万m²
建筑高度：<150m
主要功能：办公、住宅、商业、配套设施
国际竞标：第二名

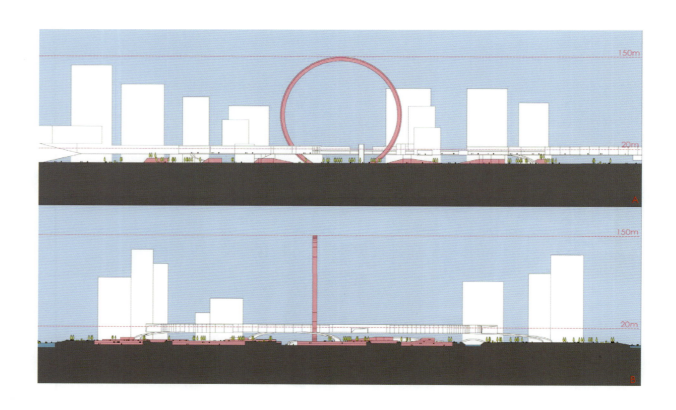

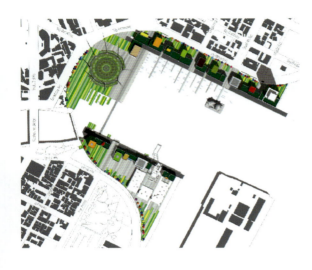

03 ARCHITECTURAL
Design of Kaohsiung Maritime Culture & Popular Music Center

The purpose of this project is for transforming the old industrial port to a nucleus of a cultural and recreational urban development.

The site is a water plaza. A large mirror of Kaohsiung city life and spirit.

We propose to cover the tensile structure standing by 8 masts and net of cables by a green living coat. The dome is not only the sky of performance hall, it's an urban landmark, a key-image of the luxurious and tropical world of Kaohsiung

We propose to develop the urban regeneration by 2 parallel lines and 1 triangle of experience through the site. 2 Parallel lines achieve the talking and the face to face of the 2 urban banks of the lover river. 1 Triangle defined by the 2 main masses of events and gathering.

We open the urban regeneration to unknown future by integrated development as hotels, apartments, SOHO, commercial, creative loft and tourism facilities. That potential could be managed by progressive process and economical increase of the value as saved resource.

台湾高雄海洋文化及流行音乐中心新建工程国际竞赛

本案是将旧工业码头改建成为城市的文化创意核心。

海岸的水广场像一面镜子，折射出高雄的市民生活和城市精神。

中心的多功能演出厅，由8根放射柱支撑的绿色穹顶建筑，形成了一处城市地标，并与其前方的水广场一同将城市氛围引向海岸。

空间组织上，在河口两岸形成平行的聚集活动场所，并通过室内外表演建筑和端头的高层建筑形成的三角形空间关系，定义了空间特征。

设计将酒店、公寓、SOHO公寓、商业、创意LOFT、旅游功能结合，为这个地块的未来留下充分的可能性，并为今后的发展保留价值不断提升的资源。

023 · PLANNING

**Client : Costruction Office, Public Works Bureau, Kaohsiung
City Gouvernment, Taiwan, R.O.C**
Location : Lingya District, Kaohsiung
Land area : 10ha
FAR : 0.71
Building area : 71000m²
Function : Theatres and other facilities
International bidding : Bidding project

客　　户：高雄市政府工务局新建工程处
位　　置：高雄市苓雅区
用地面积：10hm²
容 积 率：0.71
建筑面积：7.1万m²
主要功能：剧院、露天剧场及配套
国际竞标：竞标方案

function layout

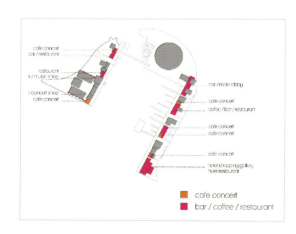
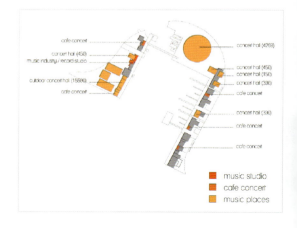
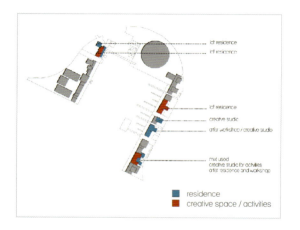
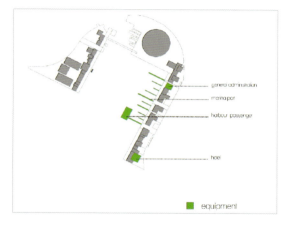
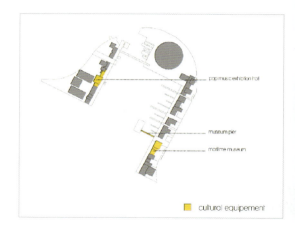

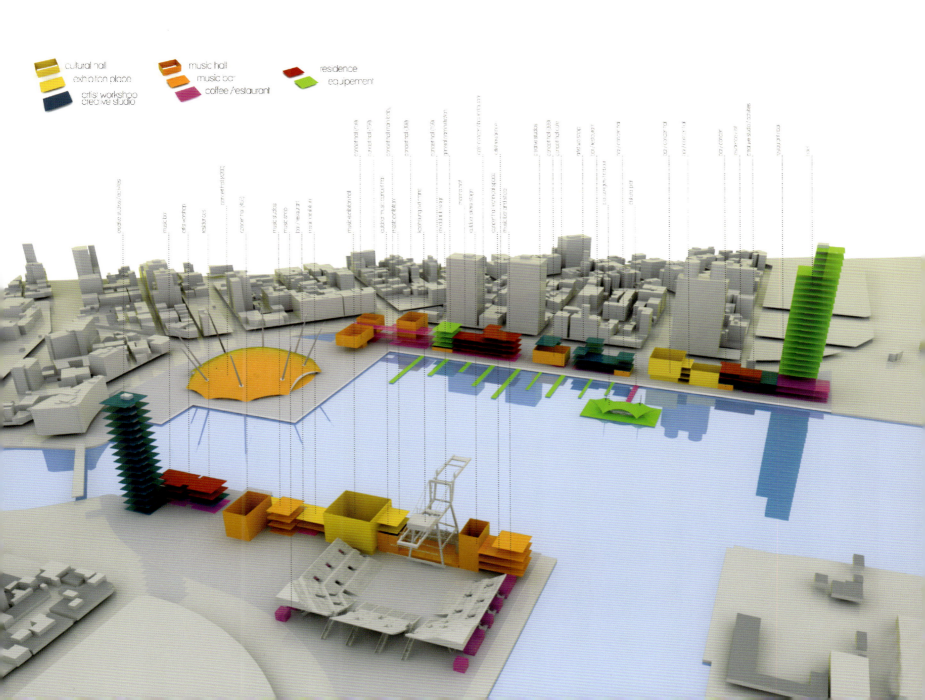

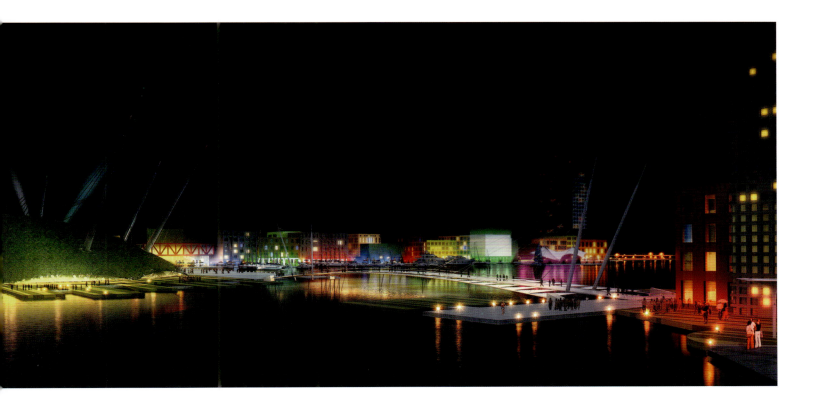

04 PLANNING
and Architecture Design of Foshan West Station

Based on the principles of high efficiency, reasonability, readability and enrichment, we want to create a landmark for Foshan through our design.

In our proposal, the building looks like a big tree, springing out from underground, spreading out in all directions to involve the two parts: underground and up-ground into the spatial development.

For creating an integrated and beautiful environment, the urban public open spaces are connected together by ecological gallery. In this way, a multi-dimensional complex attached to the station is created.

An elevated circulation system with two floors separates the arrivals and departures. A ring road is dedicated for serving the passengers and integrating the road network of south and north area, at the same time it could reinforce the accessibility of the two sides of the station.

佛山西站综合交通枢纽概念性规划建筑设计

总体以高效、合理、可读、丰富为设计原则，以树立火车站核心区标志形象为目标。

建筑源于地下，拔地而起，四面延伸，形似一棵参天大树，综合了地下、地表和地上三个层次空间的开发，形成一个城市核心。

空间景观的塑造，通过生态廊串联各个城市开放空间板块，形成依附于交通枢纽的多向度复合空间。

建立二层高架道路系统分离到达和出发交通；设置地面环状道路服务主要进出客流，整合南北片区路网，强化贯穿车站两侧的通行能力；结合广场、绿地、建筑建立二层步行系统。

Client : Foshan Planning Bureau, Foshan Railway Investment Construction Group Co., Ltd.
Location : Nanhai District, Foshan
Land area : 31ha
FAR : 0.26
Building area : 80 000m²
Height : 48m
Function : Railway station and facilities
International bidding : winning project
Cooperator : Shenzhen Urban Transport Planning Design and Research Center Co.,Ltd

客　　户：佛山市规划局、佛山市铁路投资建设集团有限公司
位　　置：佛山市南海区
用地面积：约31hm²
容 积 率：0.26
建筑面积：约8万m²
建筑高度：48m
主要功能：火车站及周边配套
国际竞标：中标方案
合作单位：深圳市城市交通规划设计研究中心有限公司

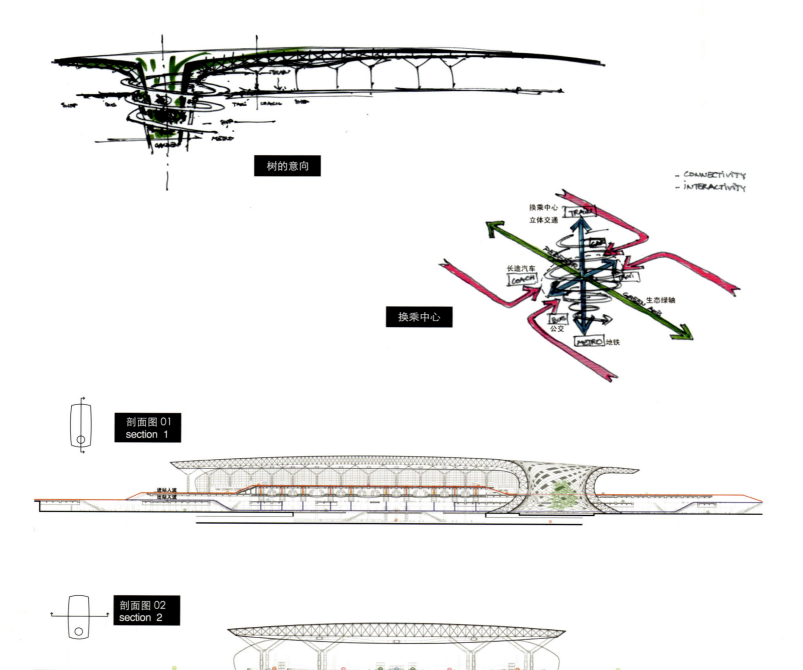

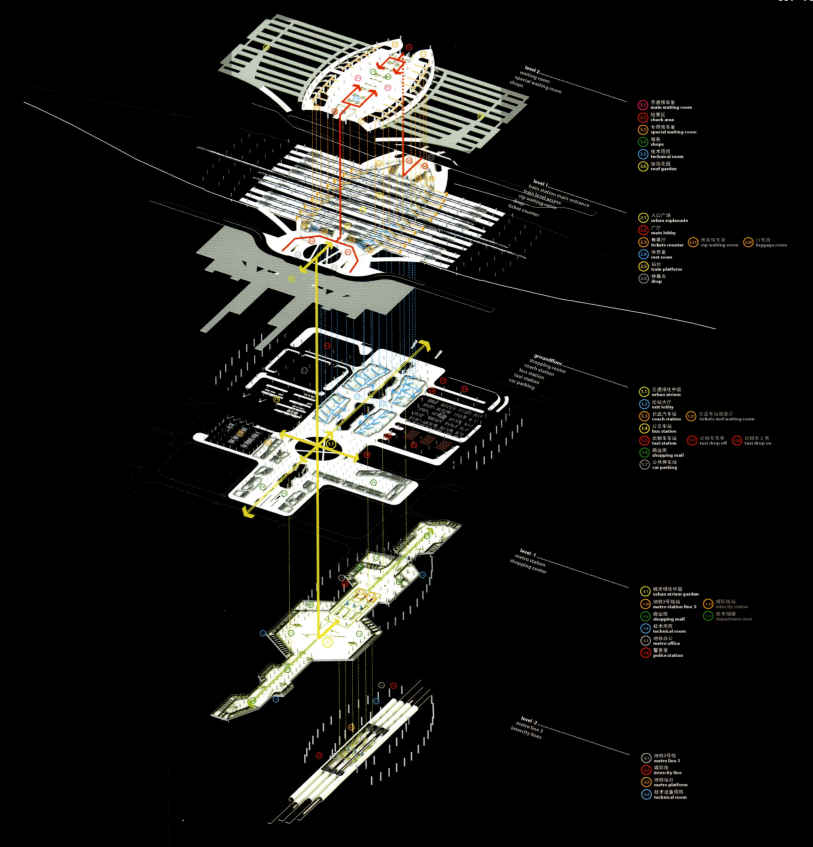

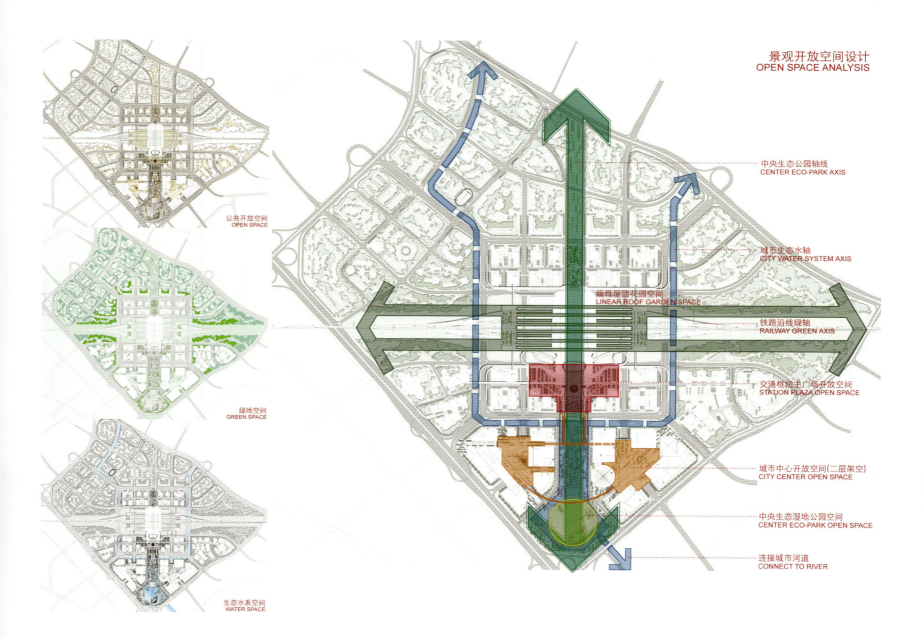

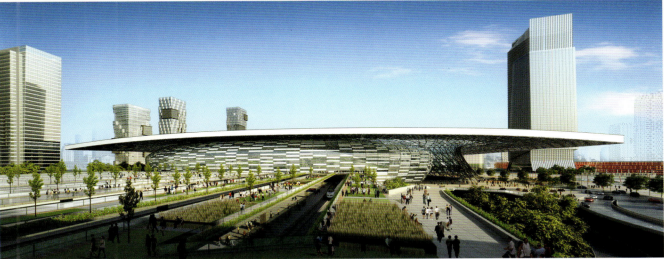

05 URBAN
Design of EBD of Jinyue District, Changchun

Strategy 1:

The concept of 'vacancy planning' has been proposed and implemented in the proposal. As the power source of urban life, the negative space is the monitoring core of government in recent years. The architectural design should meet the requirement of public space and eco-gallery. The planning departments could control and manage the vacancy space by establishing controlling plans for exterior space, as the modern cities like Singapore and Shenzhen have done.

Strategy 2:

The concept of "new urbanism" has been introduced into proposal. There are three approaches to realize it in China: creation of new towns in suburbs, regeneration of old city and urban-rural fringe zone. Jingyue EBD is the new urban district in urban-rural fringe zone. As it is close to the city, the investor could save considerable development cost, and as it is on a large scale, the designers could have a lot of choices and flexibility in the spatial distribution. New urbanism is concerned more on public space (public>private), which means that it emphasizes the harmony between people, the harmony between people and community, people's participation and sharing in community and people's cognition for physical environment. We reinforce the intercommunication, interrelation, cognition and sense of security of the people of community by adjusting the physical environment.

These two strategies are integrated together to activate the public space of this area.

长春南部新城净月西区生态商务金融中心（EBD）城市设计

策略一：

'负规划'理念的提出与贯彻。负空间作为城市生活的动力源和生命之根本，是近几年各地方政府监护的核心内容。城市实体部分设计（即建筑单体设计）应顺应虚化的公共空间与生态廊道的要求。借鉴新加坡、深圳等城市的做法，规划部门可以通过制定外部空间控制图对负空间进行更有效的控制管理，从而保证城市公共空间的品质。

策略二：

引入"新都市主义"理念，新都市主义在中国通常有三个实现途径：在郊区打造新市镇；旧城改造；城乡结合部。净月EBD属于在城乡结合部打造新城市的类型，它既可以依托城市原有的市政配套，大幅度节约开发商的开发成本，又因为规模较大，为开发商进行社区的设计提供了丰富的空间和可能性。同时新都市主义理念提倡将重点放在公共空间上，即public > private，强调人与人的和谐，人与社区的和谐，强调人们在社区中参与和共享，强调人们对社区物质环境的认知感。通过调整环境中的物质因素（景观设计的人性化）来加强人们的社会交流，从而强化社区居民的相互联系、认同感和安全感。这一点与负空间理念相互支撑，激活整个区域的公共空间生活。

035 · PLANNING

Client : Management Committee of Jingyue New District, Changchun
Location : Jingyue west district, Changchun
Land area : 319.5ha
FAR : 1.8
Building area : 5 745 000m²
Height : <180m
Function : Complex of office, commerce & apartment
International bidding : 2nd place (1st place ARATA ISOZAKI & ASSOCIATES)

客　　户：长春净月新区管委会
位　　置：长春市净月西区
用地面积：319.5hm²
容 积 率：1.8
建筑面积：574.5万m²
建筑高度：<180m
主要功能：办公、商业、住宅、配套设施
国际竞标：第二名（第一名　矶崎新工作室）

06 CONCEPTUAL
Urban Design of Huaxi Tourism Resort of Guiyang

Project Profile The conceptual urban design of Huaxi aims for establishing a frame for development and operation for the whole city. Occupied a land area of 108.6km² and a construction area of 37km², the project involves old city renewal and new city development by creating different functions, such as culture, entertainment, tourism, commerce, office, residence, university. According to the estimation, the total building area will reach 52,880,000 m².

Design Philosophy
Followed by the principles of "energy economy", "friendly environment" and "perfect combination of tourism and culture", we want to create a pleasant city integrated local natural and cultural characteristics for habitation, work, tourism and education.

花溪山水度假旅游城贵阳花溪概念性城市设计

项目概况：

本项目为"超大尺度城市运营开发"类型的概念性城市设计。其占地约108.6km²，可开发建设用地约37km²，设计内容全面涵盖文化、娱乐、旅游、商业、办公、居住、旧城改造、新城拓展、大学城等，预计至城市建设成形后，总建设开发量达5 288万m²。

设计理念：

本案以"资源节约"、"环境友好"、"旅游文化完美结合"为设计主导，结合贵阳独特的自然特征、多元的风俗文化、便利的交通条件，以第三代城市规划理念，致力于将花溪建设成为宜居、宜业、宜游、宜科研教育文化的"一河三带山水度假旅游城市"。

037 · PLANNING

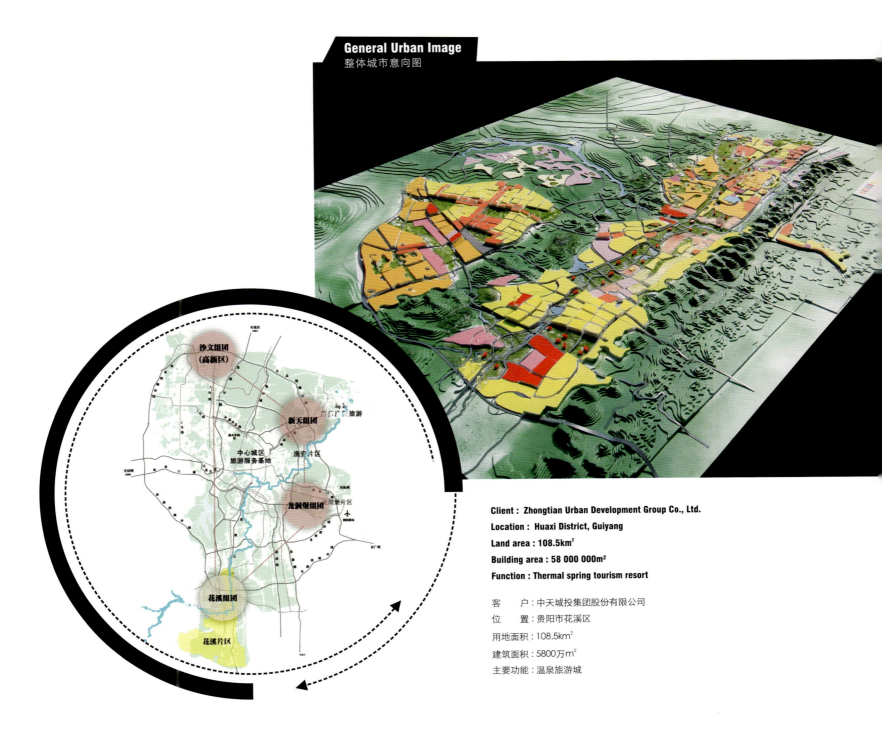

General Urban Image
整体城市意向图

Client : Zhongtian Urban Development Group Co., Ltd.
Location : Huaxi District, Guiyang
Land area : 108.5km²
Building area : 58 000 000m²
Function : Thermal spring tourism resort

客　　户：中天城投集团股份有限公司
位　　置：贵阳市花溪区
用地面积：108.5km²
建筑面积：5800万m²
主要功能：温泉旅游城

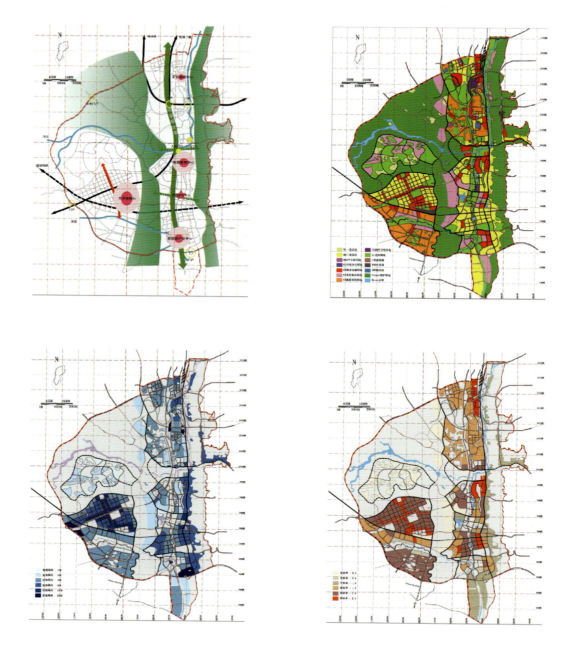

New Tourism Industry Center
新旅游产业中心

Ecological Tourism Park
生态旅游主题公园

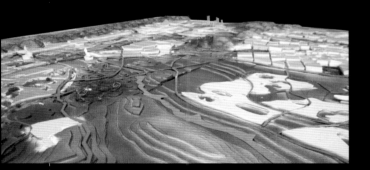

Agriculture Tourism Zone
农业观光带

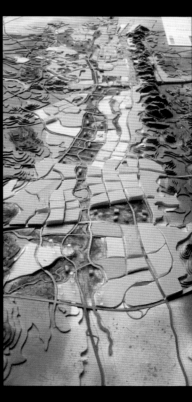

Travel Base
旅游基地中心

07 CONCEPTUAL
Planning and Architectural Design of Zone 61 of Cofco Real Estate, Bao'an District, Shenzhen

Client : COFCO Property (Group) Co., Ltd.
Location : Bao'an district, Shenzhen
Land area : 2.4ha
FAR : 2.8
Building area : 98 000m²
Height : <100m
Function : Residence, commerce and other facilities
International bidding : bidding project

客　　户：中粮地产（集团）股份有限公司
位　　置：深圳市宝安区
用地面积：2.4hm²
容 积 率：2.8
建筑面积：9.8万m²
建筑高度：<100m
主要功能：住宅、商业、配套设施
设计竞标：竞标方案

The site is located at southeast of Zone 61 with a land area of 2.42ha. Two main roads, Bao'an Road and Xi Xiang Road are lying on southwest and northeast of the site. They run through the Xixiang District and new city central area. As the extension of Shennan Boulevard, Bao'an Road is a crucial visual gallery of Shenzhen. The suburban parks on the northwest and northeast are favorable landscape resource for the project.

As the project is located on the core are of Xixiang district, we emphasize the cultural characteristic of urban public space and create an unduplicated environment by a "floating garden" and "an active commercial block".

In the proposal, the landscape is extending to the building and spreading in the air. The buildings stand in two lines along the street, with a simple and clean façade, to form a variable space and friendly landmark for the block.

中粮地产（集团）深圳宝安61区概念性规划建筑设计

Perspective
鸟瞰图

基地位于61区东南角，西南和东北侧的宝安大道和西乡大道贯穿西乡片区和新城市中心区，其中宝安大道作为深南大道的延续，是城市重要的视觉走廊。基地西北和东北侧的郊野公园提供了良好的外部景观资源。

项目位于宝安西乡片区门户位置，在开发建设的同时兼顾城市区域公共空间文化特征的突出表达，在设计中以"上浮的景观花园"、"活跃的商业街区"等理念使其特征更具不可复制性。

为使项目沿宝安大道形象面在宝安大道建筑群像中卓然独立，用地内建筑高度以分明的前、中、后三个层次呈现南低北高的天际线轮廓。主体建筑采用沿街双排布局方式，便于形成丰富多变的街区内部空间。

建筑单体引入"景观化建筑"概念，以连续折板的手法，将地表景观系统蜿蜒地引入空中伸展。住宅立面强调简约风格，以洁净、友好、开放、人性的姿态面向城市。

Illustration of function layout
功能布局意向图

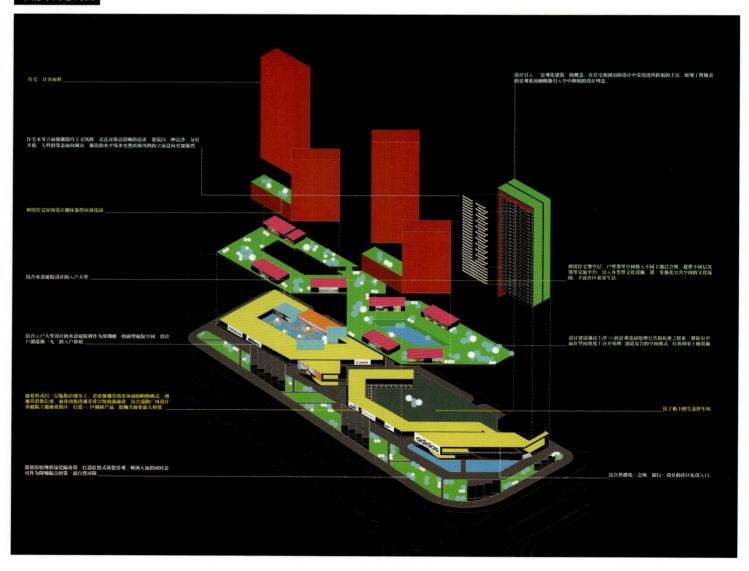

08 Conceptual
Urban Planning & Architectural Design of Xingyun Home Appliances City of Guiyang

Xingyun Home Appliances City is situated at Nanming District of Guiyang and on the end of the new latitudinal axe of the city, which connect main open spaces along the Nanming river together. Because of the location, it is an important project to promote the development of northwest area of old city.

To ensure the maximization value of land, we propose to create an urban complex with high quality in high FAR, which will be a landmark of the city. We raise the utilization rate of land by stilted floors. It provides a gathering space for people.

After design, the living condition of this area will be observably improved by means of the facilities, plantation, and activity space etc.

贵阳星云家电城项目城市设计及单体建筑概念性方案设计

星云家电城位于贵阳市南明区，结合贵阳市原有发展纵轴，将南明河沿岸重要开放空间节点串联成城市空间新的发展横轴，使纵横双轴共同展开。本案基地位于新横轴终点，是带动旧城西北片区发展更新的起点，具有承前启后的战略性地位。

由此本项目定位为"高容积率高品质的城市综合体，中心城区西部的城市地标"。设计中，鉴于用地紧张，方案提出解放地面，向高空发展的设计策略。架空首层以提供小区活动空间，并达到节约用地、提高土地利用率，适当提高容积率的目标，从而保证方案的实现。

整个项目为住户提供良好的居住品质，完善的配套设施、绿化、活动场地，彻底改善现有的居住小区建筑密度大，容积率高，居住品质较差的现况，提高居民生活品质，提升片区的整体形象。

Plot 1
地块一

Client : Guizhou Lingda Jinyue Real Estate Development Co.,Ltd
Location : Nanming District, Guiyang
Land area : 3.25ha
FAR : 12.25
Building area : 527 000m²
Height : 281m
Function : Complex of commerce、offices and residences

客　　户：贵州灵达金跃房地产开发有限公司
位　　置：贵阳市南明区
用地面积：3.25hm²
容 积 率：12.25
建筑面积：52.7万m²
建筑高度：281m
主要功能：商业、办公、住宅

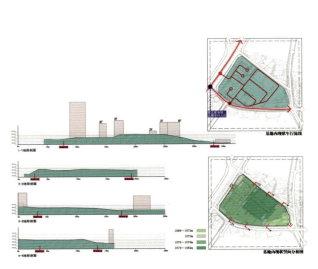
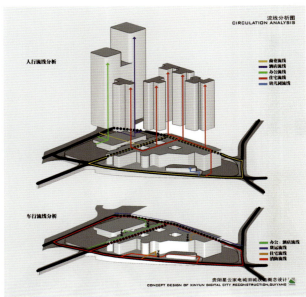
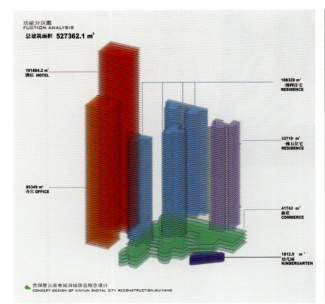

Plot 2
地块二

Client : Guizhou Lingda Jinyue Real Estate Development Co.,Ltd
Location : Nanming District, Guiyang
Land area : 2.67ha
FAR : 6.91
Building area : 250 000m²
Height : 162m
Function : Complex of commerce and residences

客　　户：贵州灵达金跃房地产开发有限公司
位　　置：贵阳市南明区
用地面积：2.67hm²
容 积 率：6.91
建筑面积：25万m²
建筑高度：162m
主要功能：商业、住宅

049 · PLANNING

09 PLANNING
and Architectural Design of Aerospace Town in Chengdu

After a series of analysis of the site, surrounding environment and urban design of the district, we proposed three conceptions as following:

'Wall and Town'

Inspired by the old city wall of Chengdu, we create a wall along the limit of site by skyline figure and façade treatment of buildings, which could visually connect the plots to make a town distinguished.

Floating Town

We use the stilts to make the buildings separated from the land, which appears as a floating town in the air. We take full advantage of Yueya Lake and City Plaza on the east to create a beautiful environment for complex.

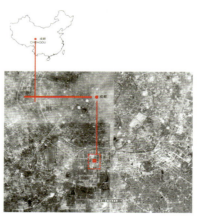

Twofold Conception

The residences are divided into two categories according to the living demands of different people. Along the limit of the site, bordering on the city, are the apartments for people who prefer a life full of energy and colors. Surrounding the Yueya Lake, are the luxury houses for people who aspire after a life with high quality.

Client : Chengdu Shenzhou Aerospace Real Estate Co., Ltd.
Location : South area of High-Tech Park of Chengdu
Land area : 4.27ha
FAR : 5.07
Building area : 300 000m²
Height : 70m
Function : Complex of residence, office and commerce
International bidding : winning project

航天成都城上城规划、建筑设计

通过对项目基地、周边环境、片区城市设计的深入分析，生成三个设计理念：

"墙与城"

设计将成都老城墙的文脉意念移植到本项目中，通过天际线的塑造及立面的处理，沿基地周边形成一道墙，从视觉上串联起整个项目的各个地块，从而将本项目打造成为一座城，形成自身独特的识别性。

悬浮之城

在建筑的底部形成一层或局部两层的架空，将上部建筑与地面形成视觉上及空间上的脱离，形成悬浮之城的概念，使月牙湖及东侧城市广场的景观多角度、全方位地渗透到本项目的内部空间中，将月牙湖的景观在水平方向上最大化的延展。

二重性

针对不同人群的生活、行为特点，将住宅分为两大类：沿项目用地周边与城市接壤的界面布置商务楼及小户型，强调城市活力，繁华、多彩；而项目内部沿月牙湖两侧布置大户型高端住宅，塑造温馨、休闲的气氛。形成入则宁静，出则繁华的中心区住宅。

客　　户：成都神州航天房地产有限公司
位　　置：成都市高新南区
用地面积：4.27hm²
容 积 率：5.07
建筑面积：30万m²
建筑高度：70m
主要功能：住宅、办公、商务楼、商业等
国际竞标：中标方案

Perspective
鸟瞰图

景观设计：
1.通过统一整合的景观设计，将公共开放的空间、小区院落空间、月牙湖、城市广场间结构成水平景观序列，并利用建筑悬浮形成的空隙，使得水平景观互相渗透。
2.竖向的景观序列设计:月牙湖——公共开放的空间——抬起的院落空间——架空层绿化——空中花园——屋顶花园。
3.纵横交叉、层次丰富的景观设计使得项目整体形象上形成花园上的城市。

诗人杜甫在《春夜喜雨》中描绘出经受了一夜春雨洗礼滋润之后的城上城，花朵红艳欲滴、饱含生机。设计以重视"城上城"的美景为目标，以月牙湖为中心将景观水平蔓延到每个地块，在竖向上以庭院绿化、空间花园、屋顶花园等渗透进各单体建筑，从而将本项目打造成"都市中的花园，花园中的社区"。

在建筑的底部形成一层或局部两层的架空，而将上部建筑与地面形成视觉上及空间上的脱窗。进而形成悬浮之城的概念。使得各小组团内的空间与月牙湖形成直接对话，并将月牙湖及东侧城市广场的景观多角度、全方位的渗透到本项目的空间内，将月牙湖的景观在水平方向上最大化的延展。

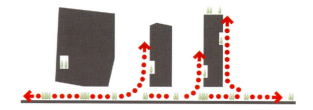 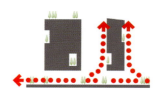

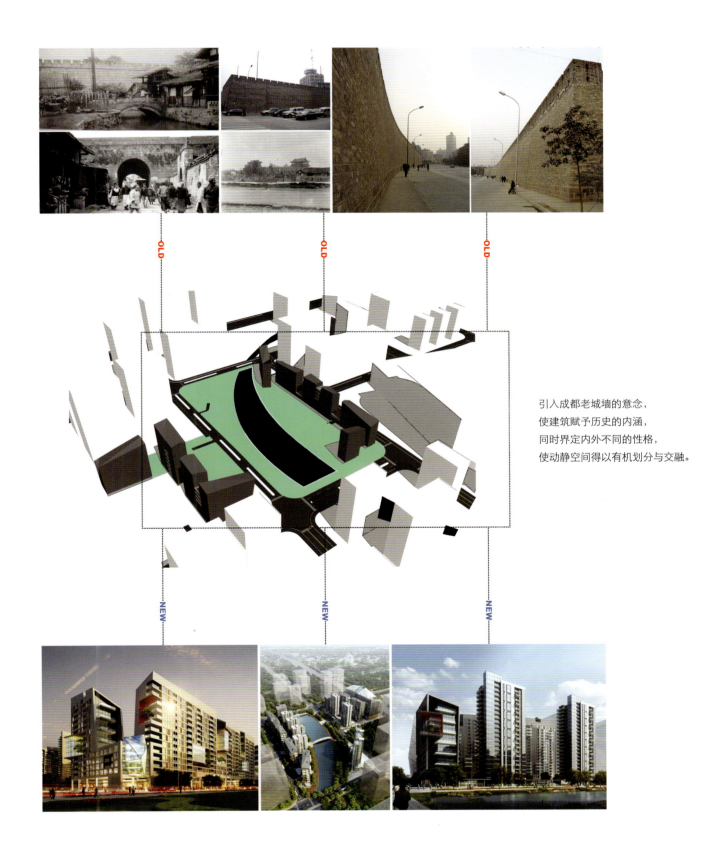

引入成都老城墙的意念,
使建筑赋予历史的内涵,
同时界定内外不同的性格,
使动静空间得以有机划分与交融。

Section 1
剖面图一

Section 2
剖面图二

Perspective
透视图

10 DETAILED
Blueprint Modification of Peninsula Residential Community in Shenzhen

The Peninsula of Shenzhen occupies a dominant place close to Shenzhen Bay. Its 15km pedestrian belt is an import part of coast belt of Shenzhen. According to existing environment conditions, we take full advantages of the ecological landscape resources of Shekou mountain and Shenzhen bay to create natural recycling and self-adjusting microcosmic environment by following the strategy of symbiosis and sustainable development. This area will be developed in high FAR but low density to ensure the functions and environment quality. To realize the value maximization of land, the towers stand facing to sea, which will be another landmark of Shenzhen.

深圳半岛城邦三四五期详细蓝图修编

项目地理位置显著,堪称深圳第一滨海门户。基地紧紧依托15km滨海步行带,是深圳滨海走廊的重要节点,依据现状环境条件,强调生态理念,最大化保护和利用蛇口山和深圳湾生态景观资源,创造自然循环与自动调节的微观环境,体现共生思想及可持续发展略。在确保基础设施及环境品质的基础上,满足合理的开发强度,采用高容低密度的开发的模式,在重要的水岸线集中建设超高层建筑,既实现了土地价值,加强了地块的标志性,也塑造了城市的形象。

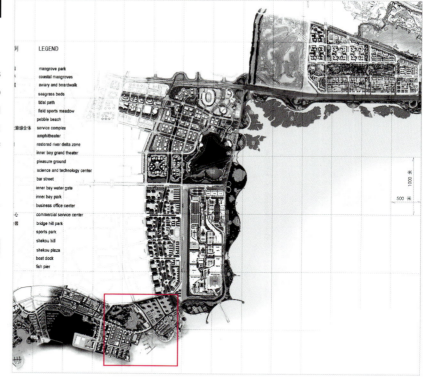

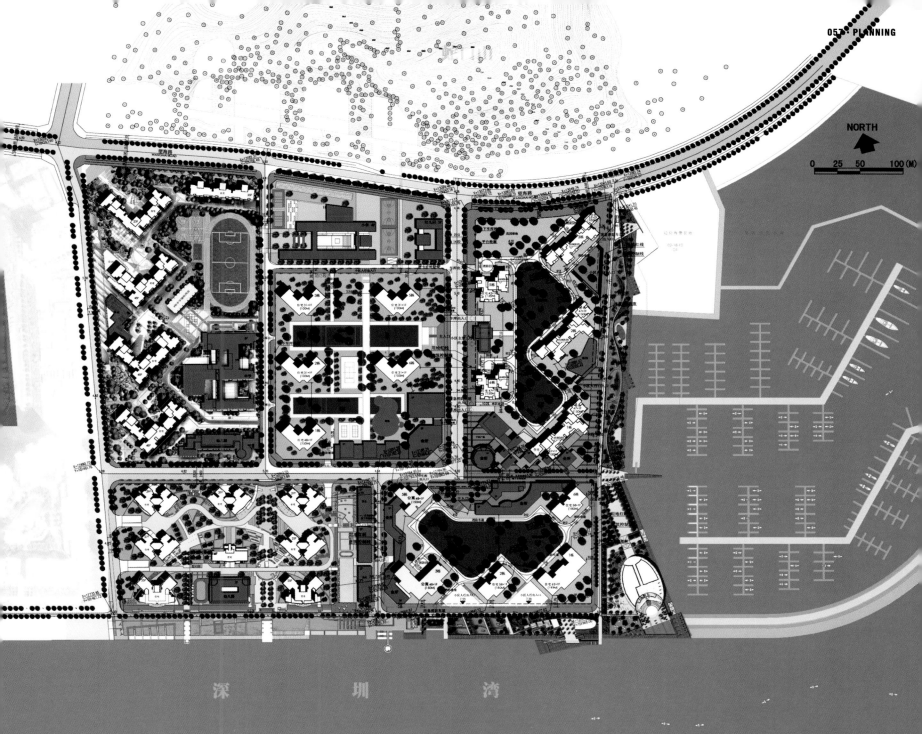

Client : Shenzhen Nanhai Yitian Real Estate Co.,Ltd.
Location : Nanshan District, Shenzhen
Land area : 26.42ha
FAR : 3.5
Building area : 930 000m²
Height : 100—200m
Function : Residence

客　户：深圳南海益田置业有限公司
位　置：深圳市南山区
用地面积：26.42hm²
容 积 率：3.5
建筑面积：93万m²
建筑高度：100~200m
主要功能：住宅

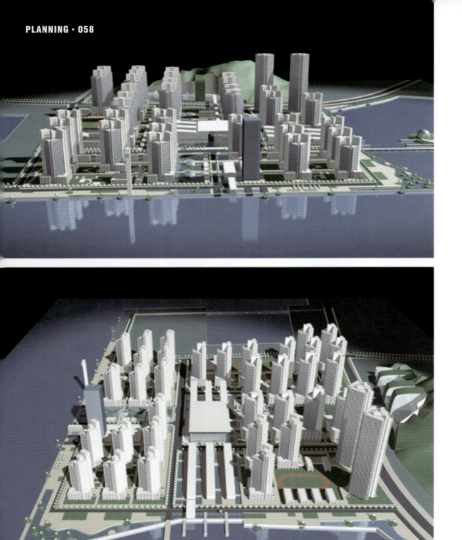

059 · PLANNING

⑪ PLANNING
and Architecture Design of Resettlement Housing of Fuzhuo CBD

As its location advantage of being close to future CBD and its unique landscape resource from City Plaza and Binjiang Park, we propose an idea of "LOHASLIFE + Park" for this project.

The essence of local living culture is implanted in the planning design in a modern way. For creating an enjoyable living environment, the designers take not only the local culture, but also the international ambiance and future development of CBD into consideration.

福州中央商务中心安置房规划建筑设计

本项目紧邻规划中的中央商务中心,极大地提高其附加值,具有显著的区位优势。南侧的城市广场和滨江公园使本项目具备独特的景观优势。根据地块以上核心价值,方案总体立意确定为"坊生活+公园"（LOHASLIFE+PARK）,从福州本地生活文脉出发,转译为现代模式化语言,形成独特的规划理念。结合中央商务区国际化的整体氛围、本地文化和地区未来发展,提供适合安置住区的全方位立体化解决方案,营造街坊邻里生活氛围,提供健康可持续发展的住区空间。

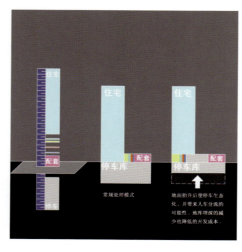

Perspective
鸟瞰图

Client : Fuzhou Urban and Rural Construction and Development Corporation
Location : CBD on north bank of Minjiang River, Fuzhou
Land area : 4.75ha
FAR : 4.1
Building area : 230 000m²
Height : <100m
Function : Residence
International bidding : winning project

客　　户：福州市城乡建设发展总公司
位　　置：福州市闽江北岸中央商务中心
用地面积：4.75hm²
容 积 率：4.1
建筑面积：23万m²
建筑高度：<100m
主要功能：住宅
设计竞标：中标方案

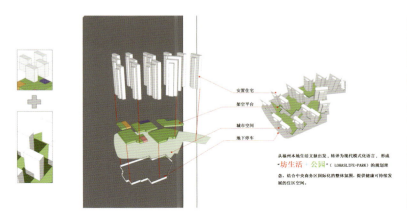

安置住宅
架空平台
城市空间
地下停车

从福州本地生活文脉出发，转译为现代模式化语言，形成
坊生活·公园"（LOHASLIFE·PARK）的规划理
念。结合中央商务区国际化的整体氛围，提供健康可持续发
展的住区空间。

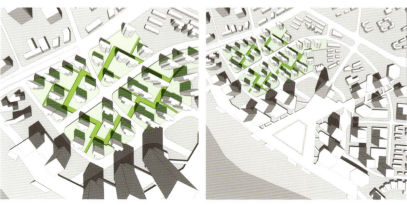

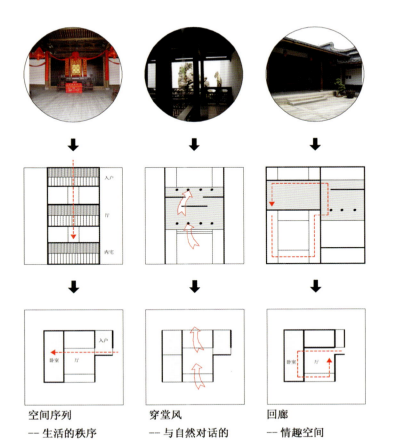

空间序列
——生活的秩序

穿堂风
——与自然对话的

回廊
——情趣空间

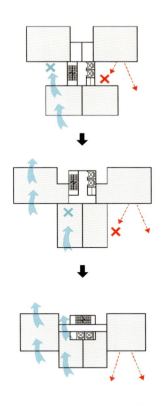

综合考虑日照，通风和视线干扰的因素之后，选择板式组合

2010
ARCHITECTURAL
建筑

01 HONGKONG-ZHUHAI-MACAO
Bridge HongKong Boundary Crossing Facilities International Design Ideas Competition

As a shinning pearl in Pearl River Delta, HongKong connects four places between two sides in China. Its Characteristic of identity, interchanging and flexibility are more and more emphasized.

HKBCF of HZMB is a heavy infrastructure of check point. We propose to rethink HKBCF as an architecture of air, a container of flexible functions and complicated circulations. In our proposal, the linear circulation substituted by centrality form smartly separated the two fields of vehicle and pedestrian. New high speed brings the high technology. The service system particularly the staff circulation as well as the emergent health service route are designed by high-tech ideas: inside of the device along the viaduct in both directions, monorail maglev train is settled so that medical cares could be sent instantly, and, the same, the patient could be smoothly evacuated.

Apart of those above, our proposal reflects a strong momentum of acceleration. It looks like all the elements rolled inside are pushed to speed up the centrifugal force. The trend that shown from the architectural form at the same time could be weighted as the epitome of HK's rapid development. The border lines and boundary crossing in the moving ramps, oblique and inclined surface on different layers, spinning around a new node which is passenger clearance building with its facilities protecting a luxurious garden at best expressing of Hong Kong urbanity.

珠海港珠澳大桥香港口岸国际概念设计竞赛

港珠澳大桥是位于珠江三角洲的一颗明珠，连接了中国的4个城市，有着重要的标识性、交换性、灵活性。

香港口岸国际检查站将是一个重要的基础设施。我们将口岸设计为空中建筑，一个装载了灵活功能和复杂流线的容器，以集中形态取代线性循环的方式，分为人车两个区域。设计提出了新科技概念，尤其是员工流线和紧急救助服务流线：在中间道的两侧双向均设有高速轻轨，用于紧急医护救助，第一时间将病患安全疏散。

设计理念致力于表现出强大的加速动力，设计内部所有的元素缠绕在一起仿佛要冲破地心引力的束缚，通过建筑的整体形态反映出香港迅猛发展的趋势。边线基础设施则设计在一个运动的斜面上，围绕着一个新中心，即乘客中转中心在不同的层面上设计斜面，通过防护性设施和茂盛的花园展现香港的城市特征。

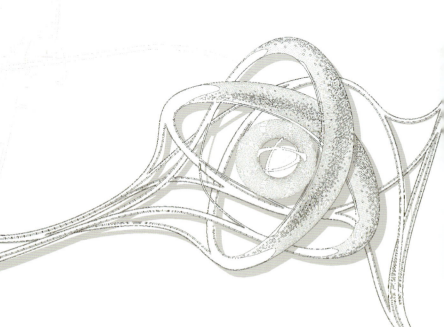

Client : Government of the HongKong Special Administrative Region
Location : Reclamation site at the waters of north-east of Hong Kong International Airport
Land area : 10ha
Building area : 400 000m²
Height : 50m
Function : Traffic hub
International bidding : bidding project

客　　户：香港特别行政区政府
位　　置：香港国际机场东北水域
用地面积：10hm²
建筑面积：40万m²
建筑高度：50m
主要功能：交通枢纽
国际竞标：竞标方案

Plan
平面图

Design Conception
设计概念图

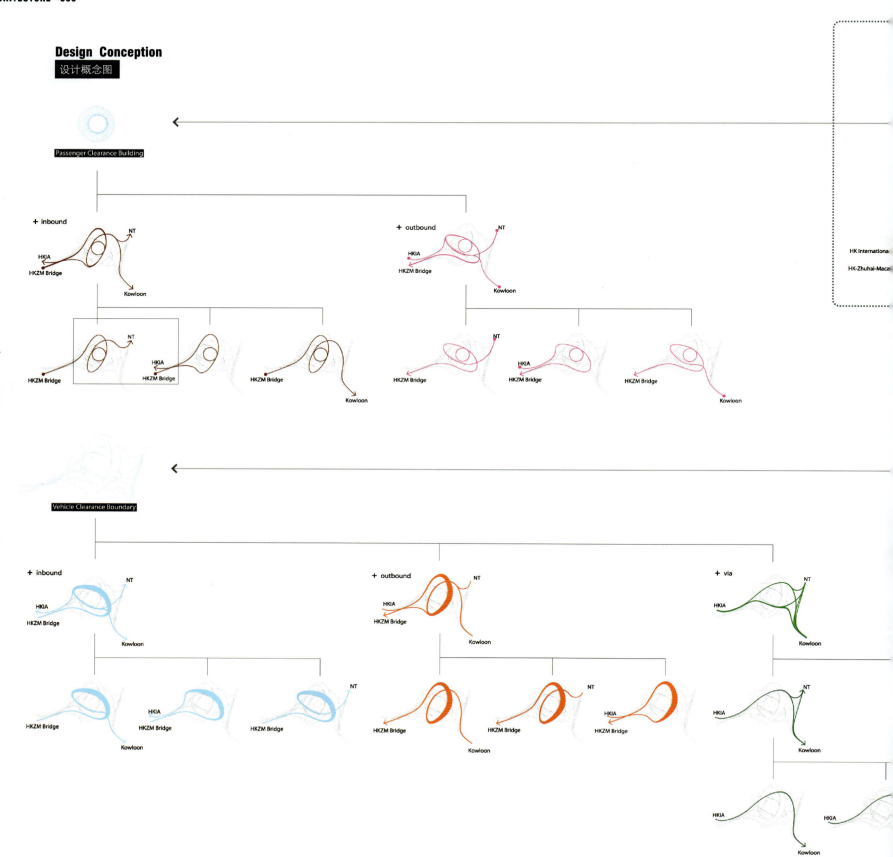

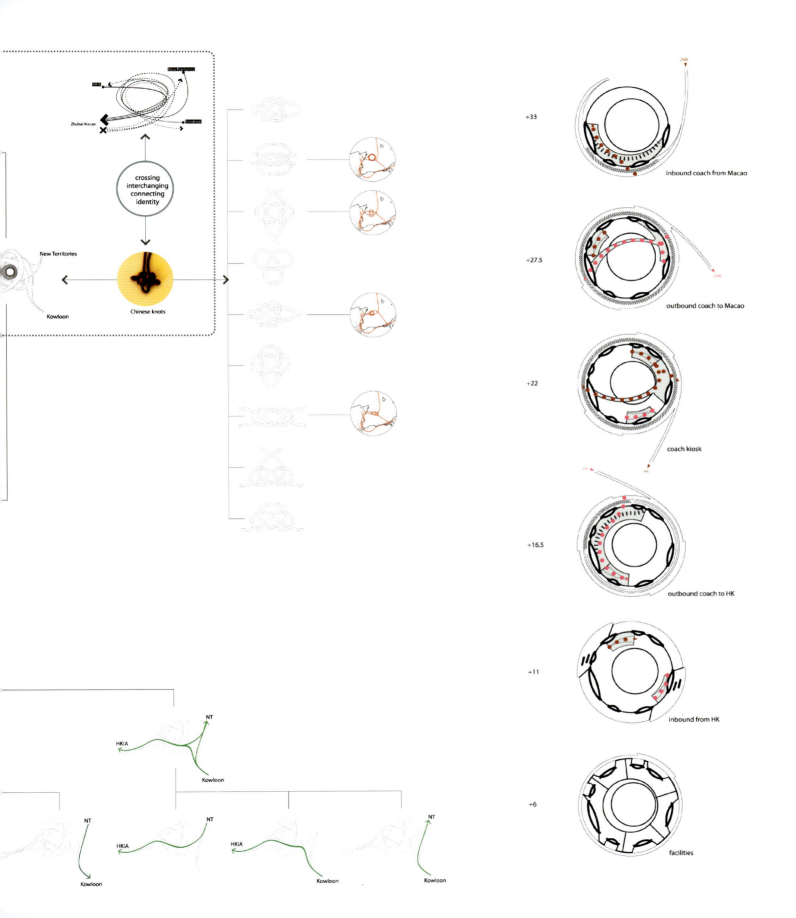

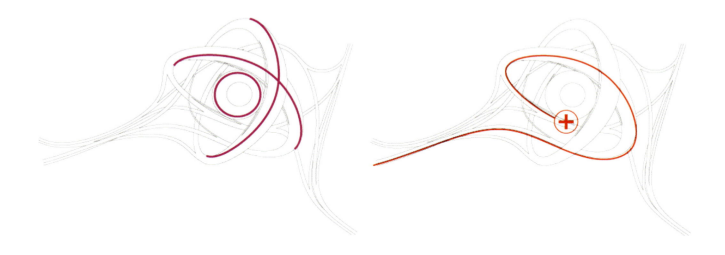

stuff circulation
員工路線

emergency monorailway circulation
緊急磁懸浮單軌路線

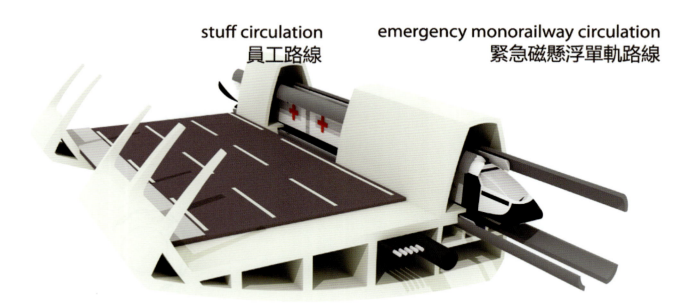

Section analysis
剖面分析图

Structure analysis
结构分析图

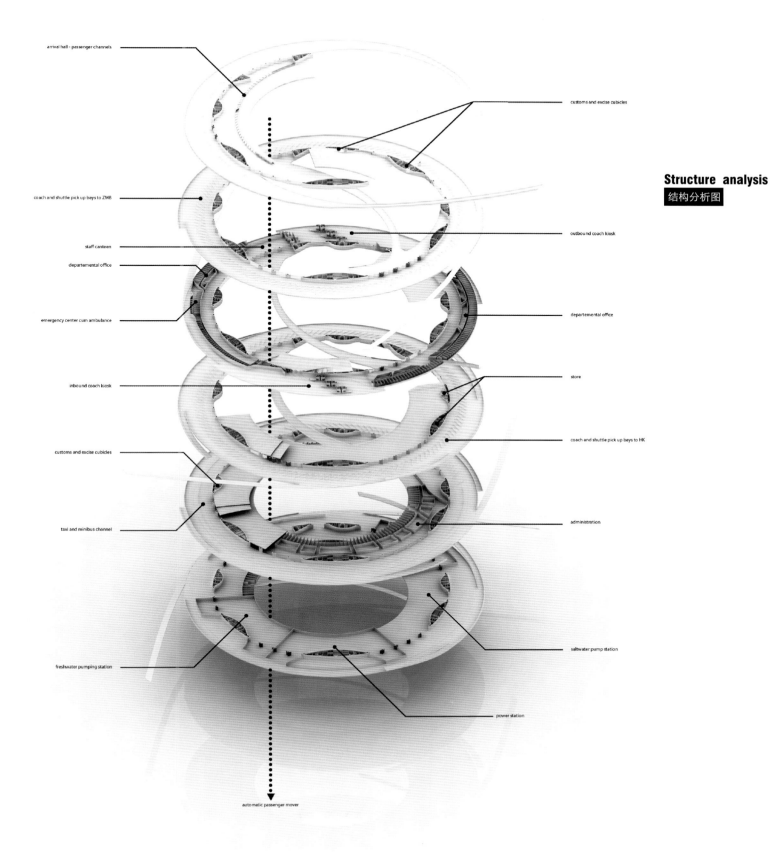

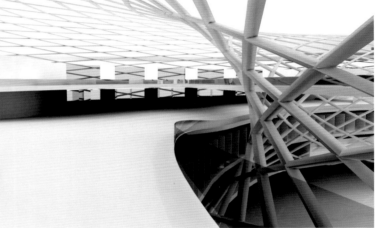

073 · ARCHITECTURE

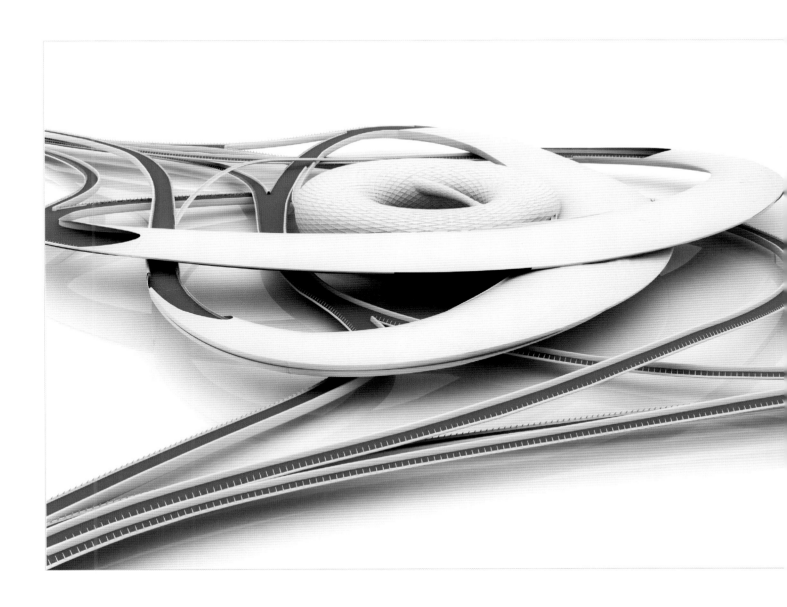

075 · ARCHITECTURE

02 ARCHITECTURAL
Design of Nanshan Culture(arts) Museum of Shenzhen

As Nanshan Culture (Arts) Museum is an important place of culture consumption of Shenzhen, its architecture design is also full of cultural and art features.

Its Form is like a mountain.

Its destiny is exchange of culture and art.

Its owner is people.

It's the embodiment of human spirit.

It isn't a big building, but inside is an ocean of art.

深圳市南山文化（美术）馆方案设计

南山文化艺术馆是深圳当代文化消费的重要一员，其诞生之初，即具有至少三重内在的意蕴：形式，取"山"，柔和葱茏，是景致人工抽象的自身镜像，故为"山之茂"；内容，载"文"，广博多元，是文化艺术交流碰撞的交易平台，故为"艺之贸"；创造，由"人"，虚怀若谷，是我们修养漫游的精神样态，故为"人之貌"。建筑虽不大不高，但得其内在，以成为一座不容小觑的人文山峦。

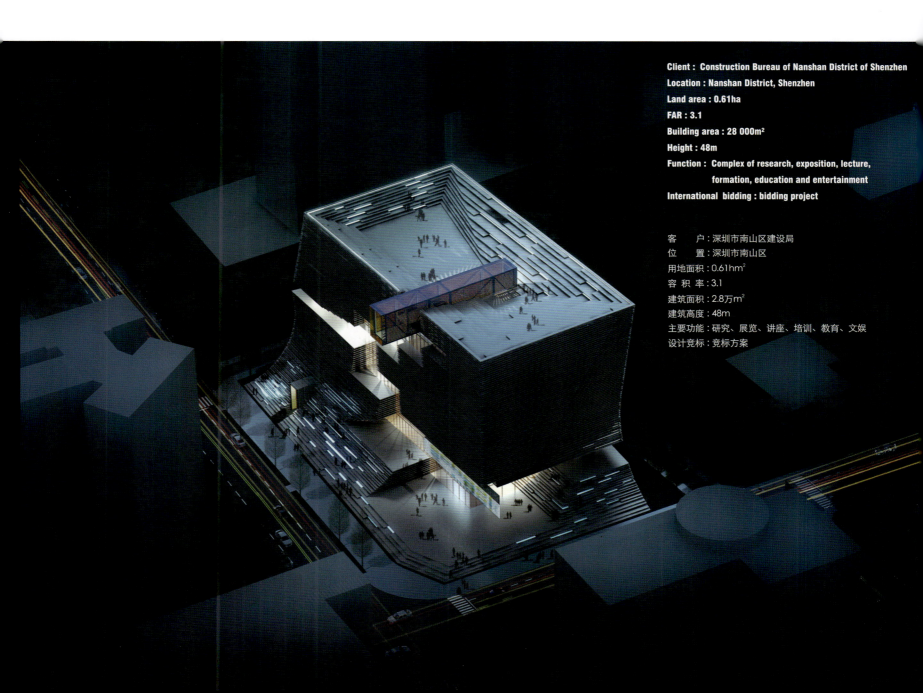

Client : Construction Bureau of Nanshan District of Shenzhen
Location : Nanshan District, Shenzhen
Land area : 0.61ha
FAR : 3.1
Building area : 28 000m²
Height : 48m
Function : Complex of research, exposition, lecture, formation, education and entertainment
International bidding : bidding project

客　　户：深圳市南山区建设局
位　　置：深圳市南山区
用地面积：0.61hm²
容 积 率：3.1
建筑面积：2.8万m²
建筑高度：48m
主要功能：研究、展览、讲座、培训、教育、文娱
设计竞标：竞标方案

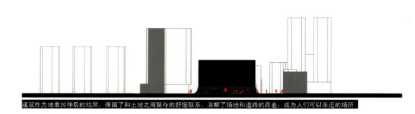
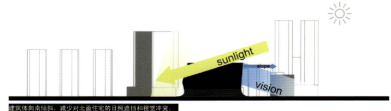
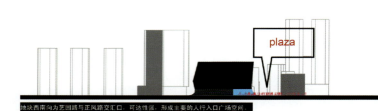
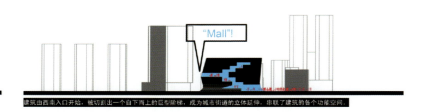
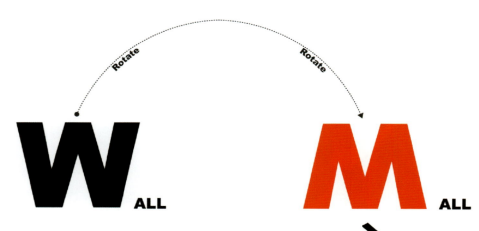

079 · ARCHITECTURE

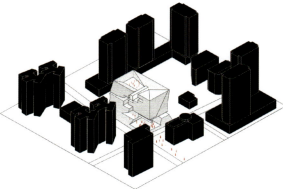
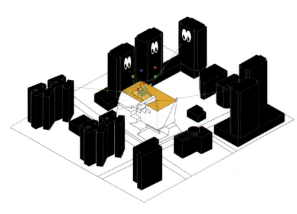
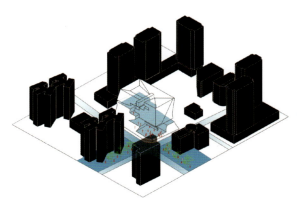

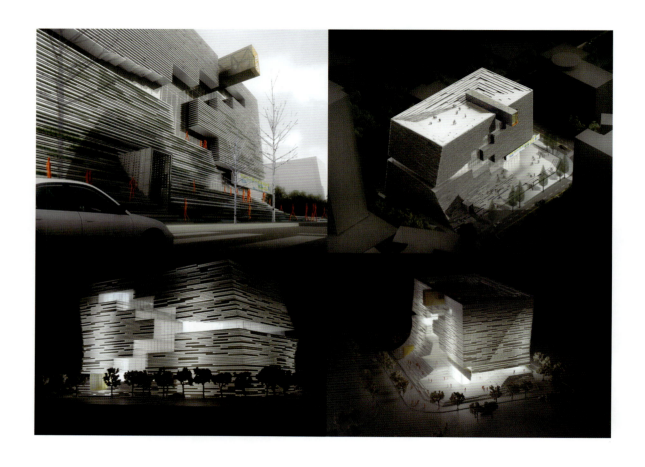

03 ARCHITACTURAL
Design of Skyworth Semiconductor Design Center of Shenzhen

After analysis and demonstrations, we applied an integrated design to unify the architecture and landscape design. Taking full advantage of the sites by planning, architecture and landscape design to create an integrated public space and to make the building correspond to the spirit of the enterprise. We chose the new material, new technology and new method to create an uncommon spatial experience and to create a new landmark for Shenzhen.

深圳创维"半导体设计中心" 建筑方案设计

通过反复的研究论证我们尝试了一种建筑景观一体化的设计，真正地把建筑当做场地来操作，广场、景观，立面使用同一元素，从规划、建筑、景观三个层面，运用创新的设计手法充分发挥用地先天优势，扬长补短，创造一体化公共空间体系。在源于企业品牌LOGO的直白表达方式与寻求内在精神契合两者之间，我们选择后者。创维的品牌定位是"睿智、创新、上升"，提案从企业精神出发，在全新的三维体系中，运用新材料、新技术、新语言，创造全新的非凡空间体验。在公众喜闻乐见的同时提升片区文化的品质，创造高新区乃至深圳市的风景和标志。

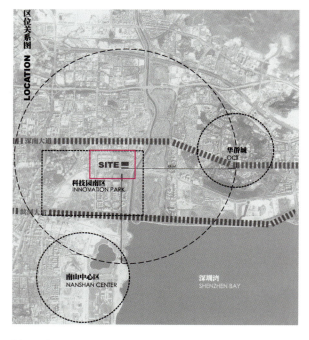

Client : Skyworth Semiconductor（Shenzhen）Co.,Ltd
Location : Nanshan District, Shenzhen
Land area : 1.7ha
FAR : 5.0
Building area : 126 000m²
Height : 100m
Function : Complex of office, conference, exposition and other facilities
International bidding : 2nd place

客　　户：深圳创维集团
位　　置：深圳市南山区
用地面积：1.7hm²
容 积 率：5.0
建筑面积：12.6万m²
建筑高度：100m
主要功能：办公、会议展览、配套设施
国际竞标：第二名

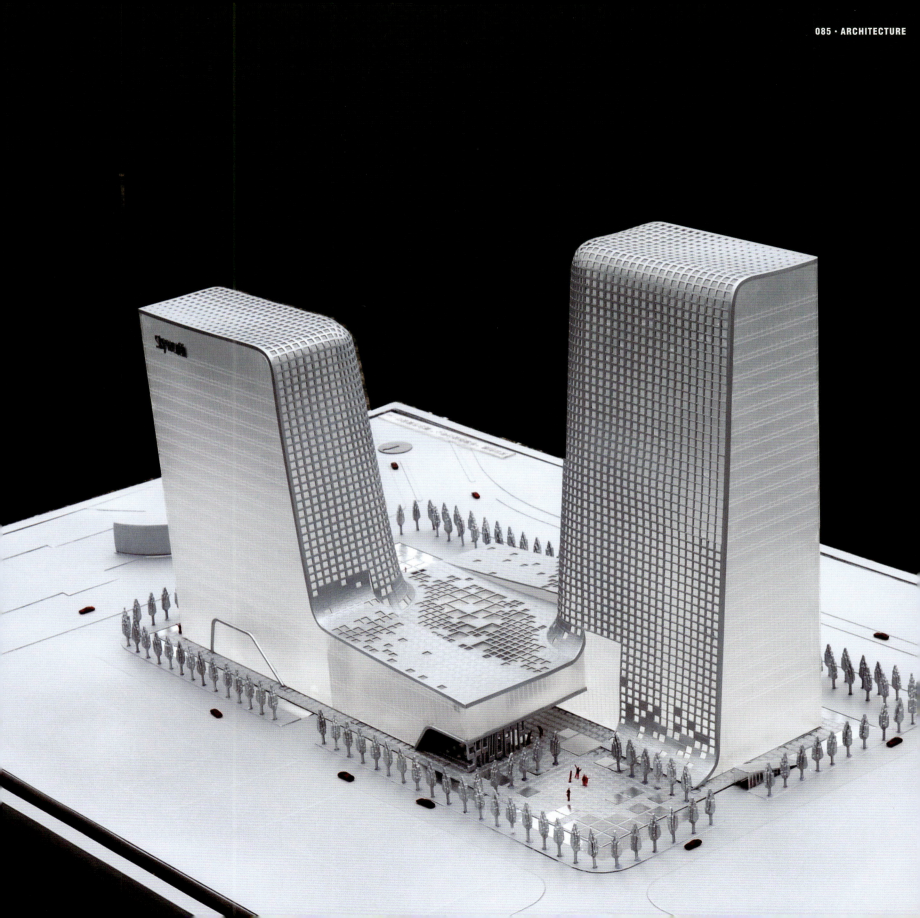

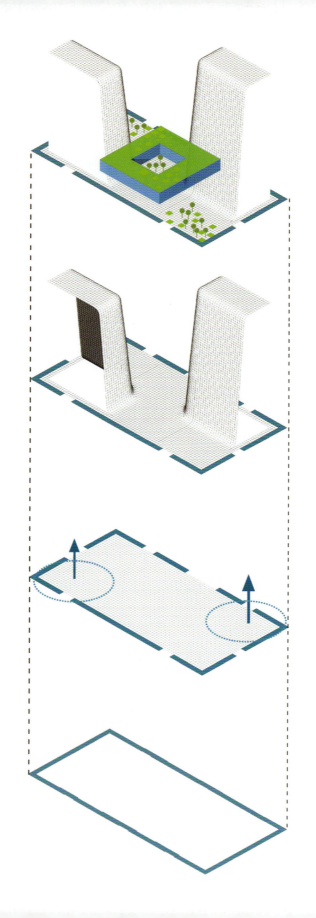

塔楼的流畅造型与集透空中庭、试验车间、会议中心、商业、餐饮为一体的商务配套设施形成双塔间独特的共享裙房。
商务配套设施顶部形成可达的上浮景观平台，有限的地表空间在空中得以再生

沿边界对角将地表向上拉伸，重点强化第三维度

在边界内部，赋予场所正交网格肌理，重新梳理水平维度

首先，引入宜人的矩形水面创造集团用地边界

景观建筑一体化的设计

087 · ARCHITECTURE

通过以上5步操作，形成一处特征性极强的城市空间，建筑景观一体化操作一气呵成，地标性与昭示性得以充分实现。

Client: Science, Industry, Trade and Information Technology Commission of Shenzhen Municipality
Location: Nanshan District, Shenzhen
Land area: 0.5ha
FAR: 14.15
Building area: 88 000m²
Height: 100m
Function: Office
International bidding: 2nd place

客　　户：深圳市科技工贸和信息化委员会
位　　置：深圳市南山区
用地面积：0.5hm²
容 积 率：14.15
建筑面积：8.8万m²
建筑高度：100m
主要功能：办公
设计竞标：第二名

04 ARCHITECTURAL
Design of United Headquarter of High-Tech Enterprises of Shenzhen

The site of United Headquarter of High-Tech Enterprises is situated on the northeast corner of reclamation VI of Shenzhen High-Tech Industrial Southern Park. It is a project invested by the government to satisfy the demands in office and R&D of enterprise headquarters.

In our proposal, the layout fully respects the existing urban design guideline. The 27-storey tower stands on the south of the site with a height of 100m. Its podium is 6 storied and with a height of 25m.

The texture of the façade is in horizontal rings, which could reduce unfavorable influence by irregular plans. The air gardens in the building make the building more graceful.

深圳市高新技术企业联合总部大厦方案设计

高新技术企业联合总部大厦位于高新区南区填海六区东北角，是由政府投资兴建的服务平台，满足企业总部办公、研发需求。

建筑布局尊重原有城市设计控制导则，塔楼位于基地南侧，在高层退线范围内置27层的直筒式塔楼，高度100m；北部裙楼沿建筑退线轮廓设计六层，高度25m。整体建筑采用环绕的水平方向肌理，以减弱不规则的建筑平面对形体带来的不利影响。同时建筑楼层分层相互错动，配合内部的空中花园，使建筑体量灵活流动，削弱大体量直筒式建筑呆板封闭的观感。

091 · ARCHITECTURE

05 ARCHITECTURAL
Design of Building Complex above Shenzhen University Station of Metro Line No.1

Now, in the modern society, we always try to make our life more convenient and more comfortable, but in the other side, when the city scale is gradually out of control, the traffic is becoming the major problem in our daily life.

The cars and rail traffic constitute a 3D circulation system for the city. In the place where they join up, we build up the transit hub.

This project is located on the traffic juncture where exist complicated and multiple traffic modes, consequently we need to create a building with multifunction. We want to well treat the relationship between the building and traffic to create a "urban cohesion".

深圳地铁一号线深大站综合体上盖物业方案设计

在现代化的城市，人们始终在追求工作生活的便捷和舒适。在失去控制的城市尺度中，城市交通成为影响人们生活最为关键的问题之一。

复合交通组成的空中、地面、地下多维度立体交通体系重申了城市脉络。当这些脉络在重要的城市节点中交汇，便形成了城市中各个区域的交通枢纽。

项目作为地铁上盖工程，正处于这样的城市交通汇聚点上，呼唤综合性的建筑功能和复杂多样的交通模式。设计希望梳理出建筑功能与城市交通脉络在项目中的联系，感受城市脉搏，创造一个"活的"城市衔接体。

Client : Shenzhen Metro Co., Ltd.
Location : Nanshan District, Shenzhen
Land area : 0.98ha
FAR : 10
Building area : 98 000m²
Height : 200m
Function : Complex of office and hotel
International bidding : bidding project

客　　户：深圳市地铁集团有限公司
位　　置：深圳市南山区
用地面积：0.98hm²
容 积 率：10
建筑面积：9.8万m²
建筑高度：200m
主要功能：办公，酒店
国际竞标：竞标方案

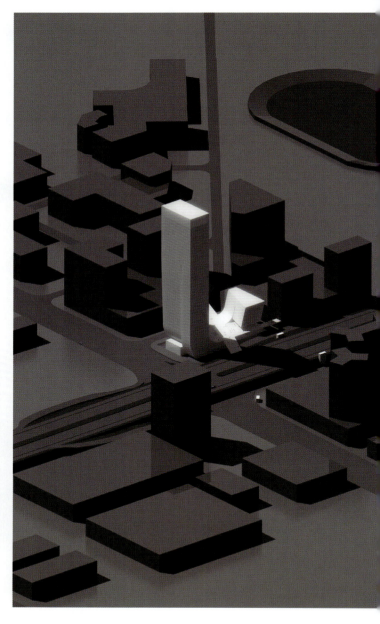

06 ARCHITECTURAL
Design of Shenzhen Mobile Tower

The design is inspired by the honeycomb-shaped structure which is the symbol of China Mobile. We hope, trough our design, to present the philosophy of "Technology changes life" and to realize the unification of the function and form.

System: We try to establish a complete architectural system by dividing the building into private space, semi-public space and public space from top to bottom. We create an ecological space through the air gardens, rain collectors, window-wall ratio and other measures. The cable floor and intelligent eco-office system are applied to demonstrate the technological strength of China Mobile.

Symbol: The special treatment of roof brings a sudden change to the skyline of the city. The building will be a pharos beside Citizen Center by its lights shinning in the night.

Communication: 3 large LED screens constitute a dynamic façade which could send the message of China Mobile to public.

Interaction: a vertical circulation system forms a special visit path, which could display the enterprise's spirit and also be convenient for the communication between employees.

深圳移动生产调度中心大厦概念性方案设计

设计灵感来源于象征移动通信的"六"面体蜂窝网络结构，设计理念在于"科技改变生活"，设计手法着重于建筑功能性与建筑形式内在逻辑美的高度统一。

系统：设计试图建立一个健全的建筑系统，公共、半公共、私密空间由下而上，连接于高效的电梯系统；以空中花园，雨水收集装置，50%的窗墙比等设计手段共营绿色生态系统，网络地板，智能化生态办公系统充分显示"中国移动"的科技实力。

标志：通过屋顶的特殊处理形成城市天际线的突变，并通过灯光塑造市民中心旁的灯塔，形成城市尺度上的标志。

传播：利用新技术，以三个大型LED屏幕形成随时间变化的动态立面，使建筑成为信息载体。

分享：在塔楼内部垂直植入线形交流空间，上升到屋顶会所，形成一条特殊参观展示路线，对外展示企业风貌，对内方便员工交流。

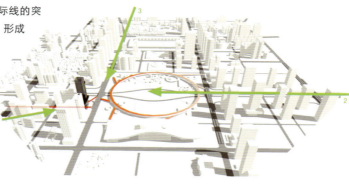

Client : China Mobile Group Guangdong Co., Ltd.
Location : Futian District, Shenzhen
Land area : 0.56ha
FAR : 13.9
Building area : 79 500m²
Height : 159.6m
Function : Office Commerce
International bidding : bidding project

客　　户：中国移动通信集团广东有限公司
位　　置：深圳市福田区
用地面积：0.56hm²
容 积 率：13.9
建筑面积：7.95万m²
建筑高度：159.6m
主要功能：办公 商业
国际竞标：竞标方案

Effect
效果图

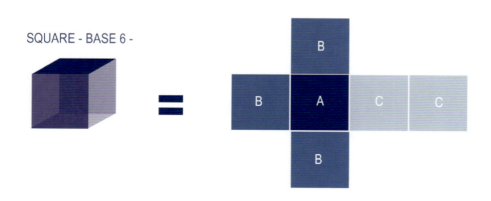
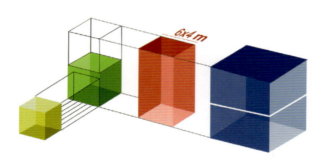
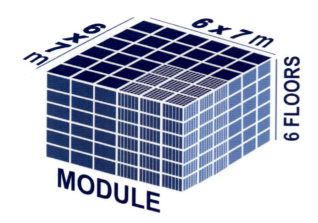
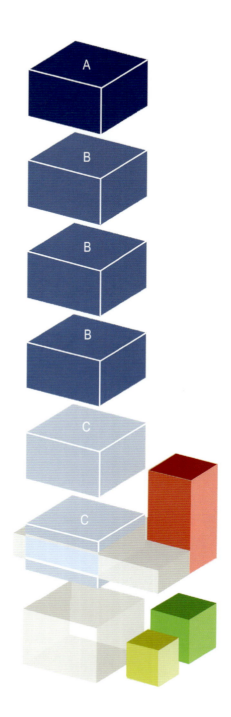

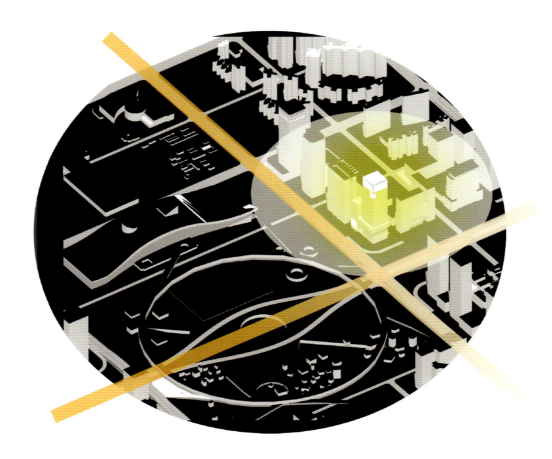

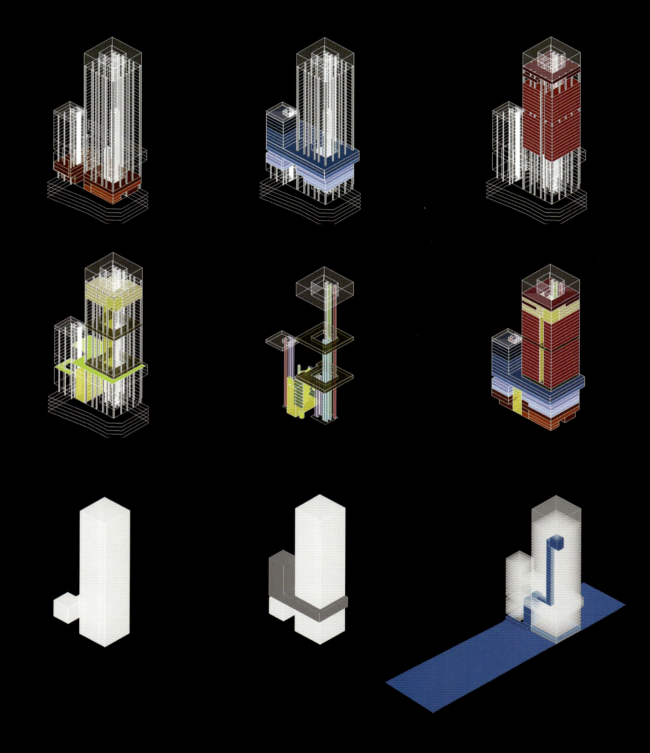

101 · **ARCHITECTURE**

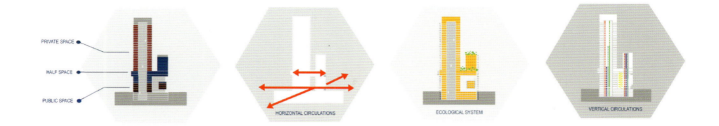

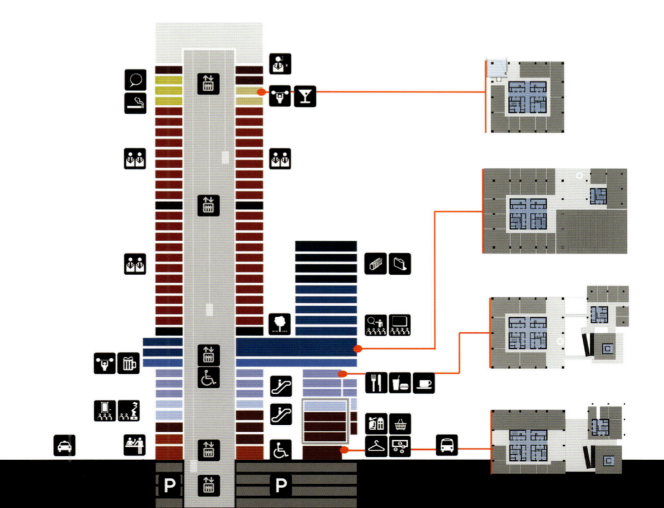

07 ARCHITECTURAL
Design of Luohu Archives Center, Shenzhen

The site is located in the city center surrounding by tall buildings, however the elegance of this project could be displayed by the gap between the buildings.

Two buildings are placed in site as two seals with different sizes and materials. For creating a govermental organization more closer to the citizens, we create 3 plazas: ecological plaza, entry plaza and culture plaza.

Inspired by chinese asymmetrical shelf, we propose a similar texture for the façade of the archive management building in south to reduce the glass area and energy consumption. For the façade of conference building, we choose the material looking like bamboos which could recreate the interlacing shadows for the inner space of this building. As the terasse of every floor are connected together, the green plantation spiral up to the top to form an air gallery for the people.

Client: Construction Bureau of Luohu District, Shenzhen
Location: Luohu District, Shenzhen
Land area: 0.66ha
FAR: 2.96
Building area: 35 000m²
Height: 60m
Function: Archives center & office
International bidding: 2nd place
Cooperator: Shenzhen Nanhua Geotechnical Engineering Co.,Ltd

深圳罗湖档案管理中心建筑设计

基地位于密集的城市组织中，周围高楼林立。设计利用周围建筑间隙形成的取景框，在城市夹缝之中，巧妙地展示本案的城市空间形象。

建筑空间的设计，形体上仿佛大小不同、材质各异的两枚印章，错落摆放于基地的南部和北侧中部。前后交错的两楼和整个周边肌理共同营造了三个公共活动空间：生态广场、入口广场和文化广场，打造出一个贴近市民的政府机构。

南部的档案管理楼模拟中国传统家具博古架纹理，减小玻璃面积，降低能耗，同时构架也有一定的遮阳作用；北部的报告楼仿佛竹简拼贴而成，"竹片"不同的偏移角度给建筑带来微妙光影变化；连续的空中阳台把地面公共空间从视觉上延伸到空中，层层阳台相连，使绿化盘旋上空，转译古典园林中的廊道，构筑成空中游廊。

客　　户：深圳市罗湖区建设局
位　　置：深圳市罗湖区
用地面积：0.66hm²
容 积 率：2.96
建筑面积：3.5万m²
建筑高度：60m
主要功能：档案管理和政府办公
设计竞标：第二名
合作单位：深圳市南华岩土工程有限公司

BIRD'S-EYE
鸟瞰图

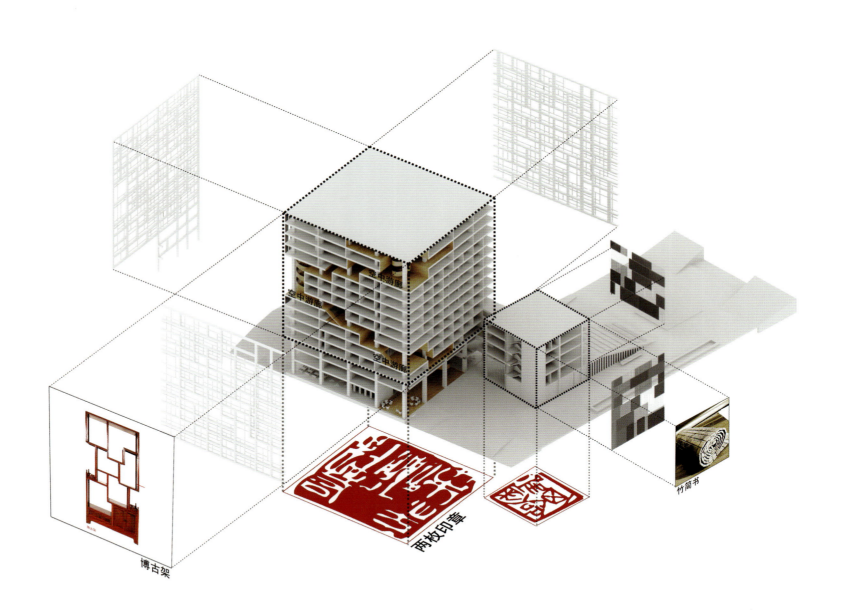

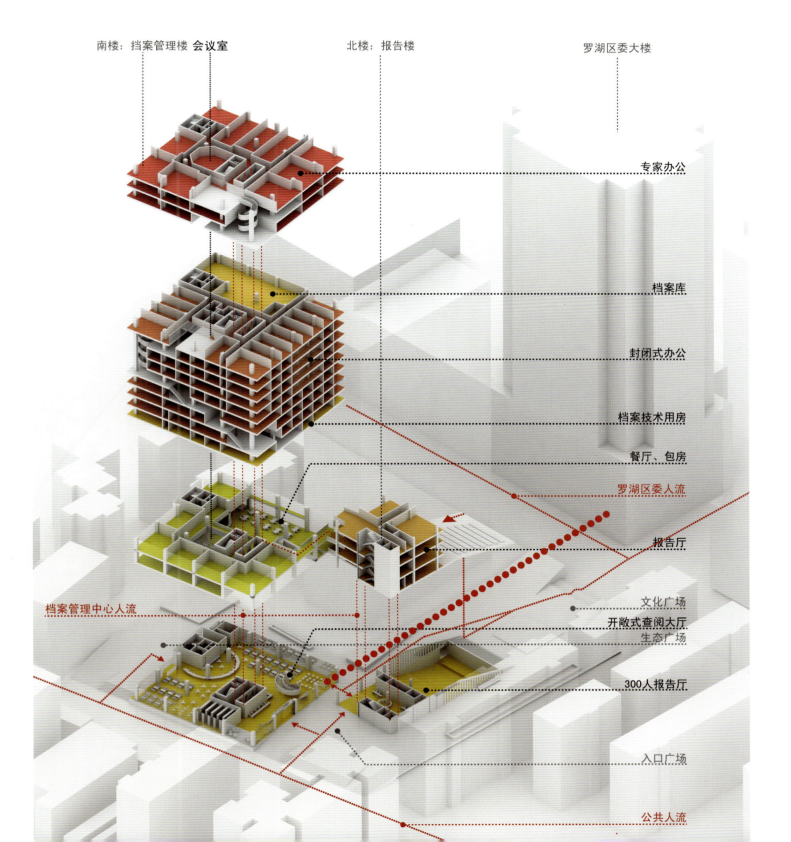

08 ARCHITECTURAL
Design of Expansion Project of Shenzhen Experimental Middle School

After platform analyses and based on the principle of openness, we propose to build a belt of ecological culture platform and connect 3 yards as well as all the functional parts through the branched distribution for creating a beautiful, harmonious and unified environment with rich cultural ambience.

After the analyses for the pedestrian flow, we organize the space planning of this area in a human oriented way. The pedestrian flows vertically intersect with the culture platform and laterally connect with the old school. The galleries under the platform are accessorial passages to the old school. The flows of culture platform vertically joined up in the new school.

We proposed a vertical circulation system around the gathering yard and place the art, music, dancing classrooms on the same floor with culture platform, so that the stands, exhibition space and temporary stage could be easily installed.

The relation between people and building could be reinforced by the visual focus.

深圳实验学校中学部扩建工程方案设计

本项目基于原有地形的分析，以"开放性"原则为指导，展开带状生态文化平台，以枝状布局串联三个庭院及纵向串联学校所有功能体，以空间编织学校肌理，释放人文气息，共同营造优美、和谐、统一的校园环境。

空间设计上以人为本，通过对人流格局的仔细分析，组织整个片区的空间规划。文化平台层和地表层流线形成垂直交错，地表层流线为水平横向渗透，连接旧校区，并于平台下层形成次要连接廊道，文化平台流线则为空中纵向贯通，分布于整个新学校。

设计中我们在人流交叉的庭院空间中设置垂直交通体系，一方面围绕庭院形成立体化的空间步行网络；另一方面尽量使文化平台层与艺术、音乐、舞蹈教室同层分布，便于临时搭建舞台、设置写生、作为运动会看台和艺术展览空间；通过进一步深化步行网络的设计，强化空中步行平台，使流线脉络分明、连接空间节点。

运用视觉焦点统领空间，增强领域感、抵达感，构成学校面向城市的展示界面。

Client : Education Bureau of Shenzhen	客　　户：深圳教育局
Location : Futian District, Shenzhen	位　　置：深圳市福田区
Land area : 0.70ha	用地面积：0.70hm²
FAR : 2.65	容 积 率：2.65
Building area : 18 400m²	建筑面积：1.84万m²
Height : 46m	建筑高度：46m
Function : High school	主要功能：高中
International bidding : bidding project	设计竞标：竞标方案

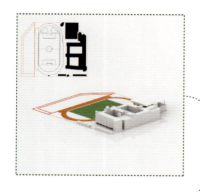

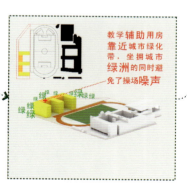

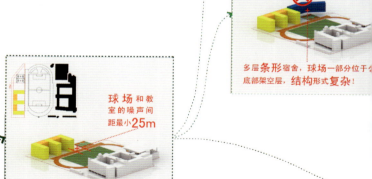

L形折板宿舍，降低高度，但它和教学楼连接生硬，影响学校城市展示面，很难处理！

Z形折板宿舍，较大的降低了高度，但它和教学楼连接仍然生硬，同时宿舍过长，影响整体立面效果

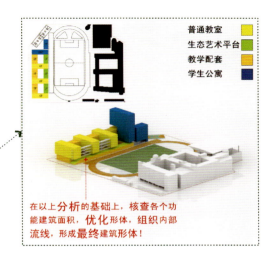

普通教室
生态艺术平台
教学配套
学生公寓

在以上分析的基础上，核查各个功能建筑面积，优化形体，组织内部流线，形成最终建筑形体！

高层板式宿舍，南北通透。球场位于宿舍后地表，不影响其结构！同时可利用宿舍屏蔽球场对教学楼的噪声！

另一个方向的L形折板宿舍，降低了高度，并和教学楼脱离，但整体形体依然平淡

增加L型一翼高度，打造标志构筑；另一翼降低到教学楼高度，建筑轮廓错落有秩，跌宕起伏，富有韵律！

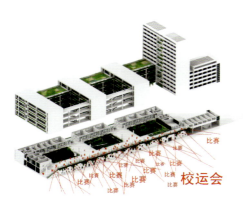
校运会

话剧节

科技月

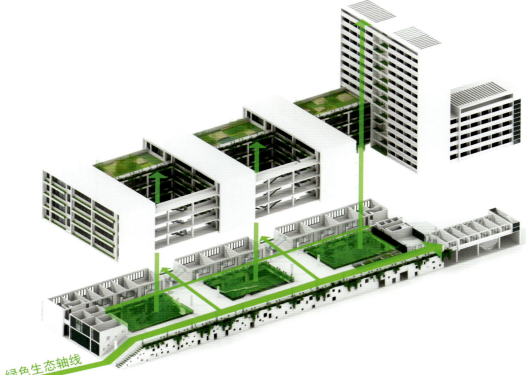
绿色生态轴线

艺术课

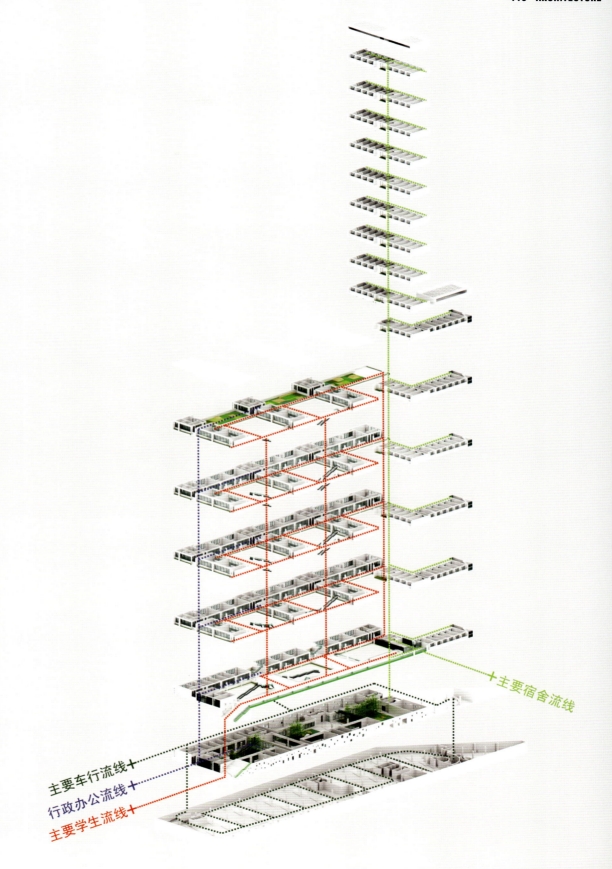

09 ARCHITECTURAL
Design of VC & PE Tower of Shenzhen

For integrating the tower with the surrounding environment and the city, our proposal makes full use of the southeast part of the tower.

In our proposal, the typical floors are divided into 5 offices with 300m², which could be combined flexibly to satisfy the demands of different enterprises.

In order to perform the social responsibilities, the concepts of 'energy conservation' and 'sustainable development' are embodied in our proposal through different measures, such as selection of façade materials and energy conservation equipments.

Client : Science, Industry, Trade and Information Technology Commission of Shenzhen Municipality
Location : New & High-Tech District of Shenzhen
Land area : 0.52ha
FAR : 12.64
Building area : 80 600m²
Height : 150m
Function : Office tower (class A)
International bidding : bidding project

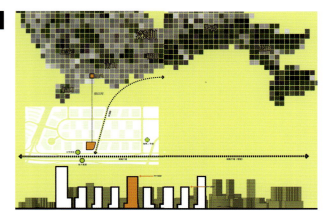

深圳创业投资大厦建筑设计

设计通过不同方式来强调并最大化合理利用建筑东南面，使建筑与城市对话，与环境共融。

办公空间高效灵活，办公标准层划分为5个约300m²的办公空间，各空间可灵活合并，以适应各类型企业入驻。

以建筑传递可持续发展的节能理念，体现社会责任感。

建筑设计注重通过立面用材、立面遮阳、节能设备选型各项环保手段，以保证相同的室内环境条件下，全年总能耗大量减少。

客　　户：深圳市科技工贸和信息化委员会
位　　置：深圳市高新区
用地面积：0.52hm²
容 积 率：12.64
建筑面积：8.06万m²
建筑高度：150m
主要功能：超高层甲级办公楼
设计竞标：竞标方案

117 · ARCHITECTURE

10 ARCHITECTURAL
Design of Shenzhen Adolescent Activities Center

Happy Together Club on the northeast of Lichi Park and at the interaction of Hongli Road and Hongling Road used to be the most popular place for the first group of immigrants of Shenzhen. It is an important part in the history of Shenzhen and it will stand out vividly in the memory of Shenzhen people.

The construction of two metro lines provides an opportunity of regeneration for this place.

The Adolescent Activities Center will be a creation focus, an energic point and a landmark of Shenzhen.

There has been, and will always be a happy land with music, poesies, dramas…

深圳市青少年活动中心改扩建方案设计

第一代深圳人和他们的孩子，都对荔枝公园东北角--红荔路红岭路交叉口的"大家乐"保留着鲜活的记忆。这里不仅是曾经的潮流，更是深圳城市重要的一部分，"大家乐"以富于教育性和创意的实践社会公众精神，创造这个城市的社会和文化精神。

两条地铁线的新建提供了契机，得以改造深圳这一块最著名最具活力的地点，向先行者致敬。这块在环境上、精神上深入人心的场所有待获得新生。

重新激活--通过生态建造和运营过程,通过持续的表演、活动、城市事件令青少年活动中心成为创意焦点、成为新交通体的活力节点，成为地标建筑，成为深圳创意的标志，成为可持续发展的深圳映像……

同时保留这块乐土，这曾经充满音乐、诗歌、戏剧……明天仍将继续，城市不仅仅是市民市场行为的产物，还表达着他们共同生活的快乐。

Client : Construction Works Bureau of Shenzhen
Shenzhen Adolescent Activities Center
Location : Futian District, Shenzhen
Land area : 2.8ha
FAR : 2.25
Building area : 63 000m²
Height : 50m
Function : Complex of theatre, formation, exposition and conference
International bidding : bidding project

客　　户：深圳市建筑工务署/深圳市青少年活动中心
位　　置：深圳市福田区
用地面积：2.8hm²
容 积 率：2.25
建筑面积：6.3万m²
建筑高度：50m
主要功能：剧场、活动、培训、展览、会议
国际竞标：竞标方案

Landscape Conception

景观概念图

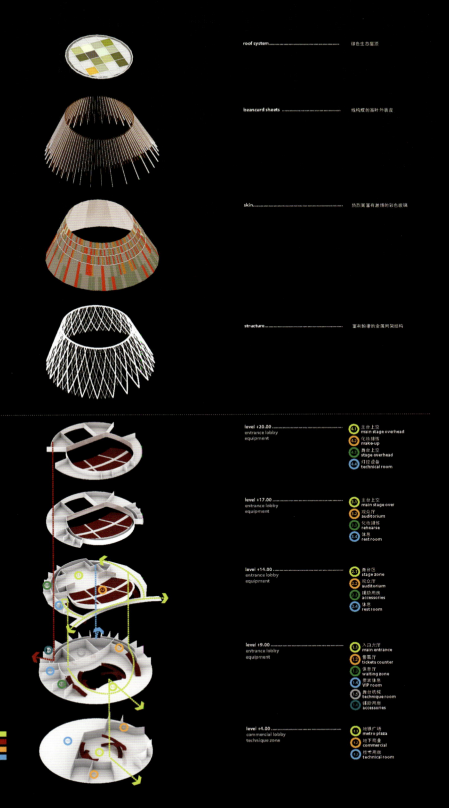

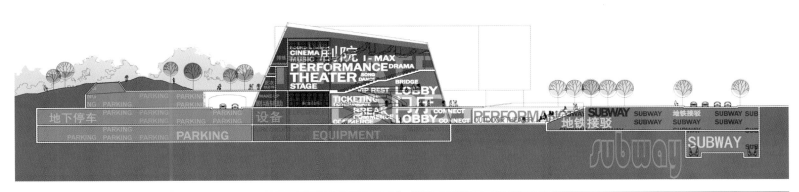

位于城市核心区域的建筑应表达的核心价值观,创新是深圳这座设计之都的灵魂,深圳的新建筑应当体现这种创新动力.

⑪ ARCHITECTURAL
Design of Shenzhen Dinghe International Tower

Situated in the core are of Futian CBD of Shenzhen, Dinghe Tower is a building of Class A offices. As its dominant location, it should well present the core value and the spirit of Shenzhen, namely, innovation.
We paid a great attention to the relationships between building and city, functions and substance to create an ecological office building with high quality.
We chose the Two Core layout for this building to realize the high FAR. A green platform was placed on the west to provide a comfortable and ecological communication place for the people. The disposition in block way well organized the circulation and made this building integrated with the city.

深圳市鼎和国际大厦方案设计

深圳鼎和大厦位于福田CBD核心区域,建设内容为甲A级超高层写字楼。设计认为位于核心区域的建筑应充分反映深圳作为设计之都的核心价值观,集中表现于对深圳城市精神的理解"创新"的表达上。
由此设计从城市关系、功能与实质、生态办公几个方面入手展开设计。
全盘考虑以上切入点,设计选择了双筒布局获得罕见使用率;西向植入绿化交流平台创造出非常规的人性化、生态化的办公空间;顺应街区布局方式以导引、整合多类流线,从而回应城市关系。

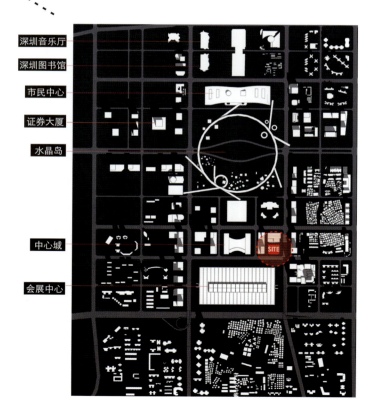

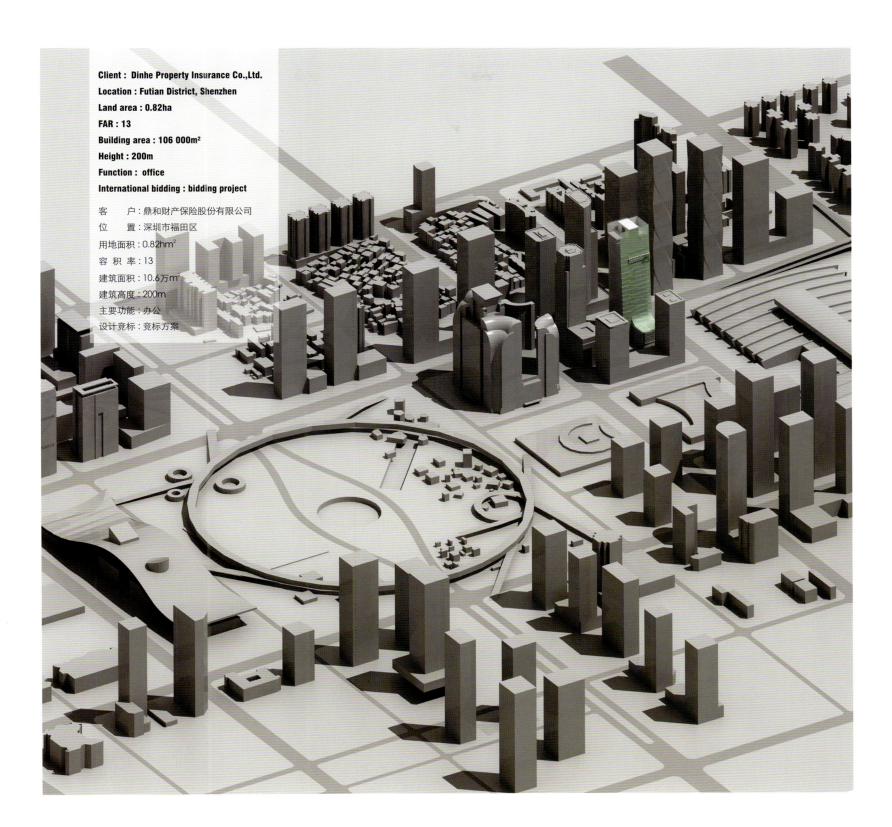

Client : Dinhe Property Insurance Co.,Ltd.
Location : Futian District, Shenzhen
Land area : 0.82ha
FAR : 13
Building area : 106 000m²
Height : 200m
Function : office
International bidding : bidding project

客　　户：鼎和财产保险股份有限公司
位　　置：深圳市福田区
用地面积：0.82hm²
容　积　率：13
建筑面积：10.6万m²
建筑高度：200m
主要功能：办公
设计竞标：竞标方案

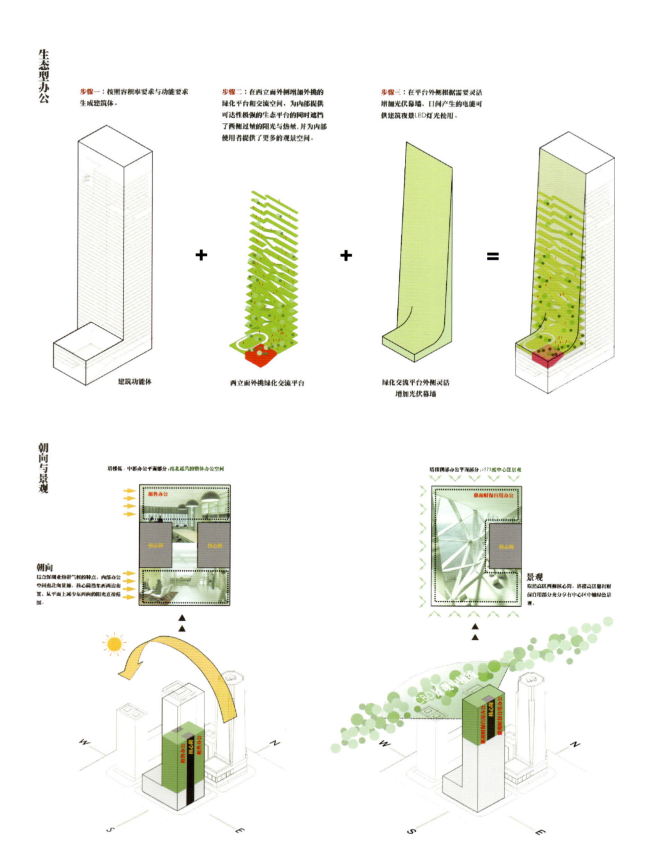

ARCHITECTURAL
Design of Resettlement Housing of Jiangong Village, Nanshan District, Shenzhen

Based on the concept of 'large community and small yards', we create an open and shared community by roads, plaza, yards and buildings. We hope we could provide spice for people's life and encourage the communication between the neighborhoods through our design. The public green land is located in the center and also on the main axis of the community, which connects with the public space of other phases by the subaxes. Although the residential groups are integrated together through the yards and other open space, their privacy is well assured.

深圳南山建工村保障性住房建设规划建筑设计

"大社区，小院落" 开放与共享，绿色与健康，诠释现代院景生活

在充分尊重城市肌理的基础上，融合传统院落布局特点及现代开放街区规划思想，以街区为源、以院落为本，将街区文化和院落内涵真正地融合在一起形成"大社区，小院落"的格局。

设计中充分运用街道广场、庭院、建筑四个载体来构建开放与共享的社区氛围，营造富有人性化、充满情趣的生活方式，促进邻里街坊的交流与互动，增进邻里关系，积极倡导绿色与健康的社区生活。

Client : Bureau of Housing & Construction of Shenzhen Municipality
Location : Nanshan District, Shenzhen
Land area : 22.42ha
FAR : 1.67
Building area : 470 000m²
Height : 60m
Function : Residence
International bidding : bidding project

客　　户：深圳市住房和建设局
位　　置：深圳市南山区
用地面积：22.42hm²
容 积 率：1.67
建筑面积：47万m²
建筑高度：60m
主要功能：住宅
设计竞标：竞标方案

ARCHITECTURAL
Design of Nanhai Middle School of Shenzhen

In this proposal, we create a 'green nursery' for school and city by 'growth gallery'. It is an ecologic and organic gallery designed with urban and social characteristics. It is a communication platform for teachers and students, which could be a miniature of society.

'Growth gallery', main circulation axe, connects different functions together and provides communication space for teachers and students.

All the functions are distributed in a perfect order, which could be beneficial to the nurtured of responsibility.

The general layout is integrated with the city to keep the unity of urban space.

Public space such as playground and green yards are provided to meet the activity demands of teachers and students.

A stilt floor is constructed for the classroom building to get natural ventilation.

Different functions are connected together by the 'growth gallery'.

深圳南海中学建设方案设计

设计以具备交流性、社会性、城市性、生态性、保护性、有机性等六大特点的"生长廊",重新诠释了"学校—城市里绿色的苗圃",从而营造一个联系紧密的场所、一个为师生提供交流机会的微缩社会。

生长廊：

构建主要交通轴线,串联学校里各种功能,赋予多样活动内容,形成富有学习氛围的交流、活动走廊,使智慧在此生长；

营造鲜明的动静分区,各类活动秩序井然,使社会责任感在此生长；

顺应城市格局排布建筑,保持连续的城市街道空间,使城市在此生长；

提供高品质的运动场、庭院等绿色空间,以保护师生日常活动安全,使新生命在此生长；

教学楼采用建筑首层架空的方式形成自然通风廊道,结合绿色技术的应用,使绿色在此生长；

令各功能通过更紧密的联系串接,有机地构成一个建筑群体,使建筑在此生长。

Client :	Engineering Affairs Bureau of Nan Shan District, Shenzhen
Location :	Nanshan District, Shenzhen
Land area :	2.4ha
FAR :	0.7
Building area :	17 000m²
Height :	24m
Function :	junior middle school
International bidding :	winning project

客　　　户：深圳市南山区建筑工务局
位　　　置：深圳市南山区
用地面积：2.4hm²
容　积　率：0.7
建筑面积：1.7万m²
建筑高度：24m
主要功能：初级中学
设计竞标：中标方案

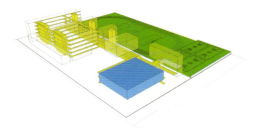

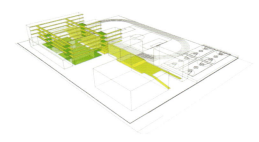

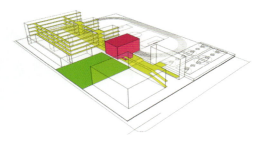

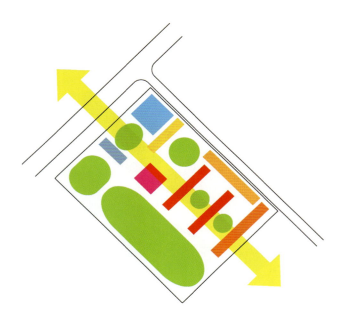

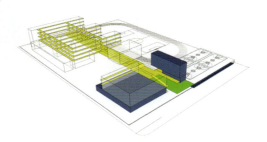

14 CONCEPTUAL
Planning and Architectural Design of Shum Yip · Bordeaux Royal Manor
(Living Service Area of Dongyuan County of Heyuan, Guangdong)

Inspired by the elegant manors and towns in Bordeaux, Pairs, Deauville, Biarritz and Nice, we proposed to create an honorable community and a high quality life style for people.

From the functional organization, we proposed a layout of "one park, two rings, three points and four districts". Taking full advantage of the river and the central green core, we create an area with the highest quality.

We try to create an ecological and convenient lifestyle by the central landscape and distribution of public service facilities. The inner circulation system is designed to be efficient, so that this project could be flexibly constructed in different phases.

河源深业东江波尔多皇家庄园规划建筑设计

规划主题源于法国优雅庄园及小镇：波尔多、巴黎、多维耶、比亚利兹、尼斯，力创一个空间明晰的环境，一个尊贵氛围的场所，一种尊享品质的生活方式。

规划布局从功能组织出发，形成"一园、两环、三点、四片区"的规划结构。利用江景和"中心绿核"，形成最高品质区域，并在各组团中心利用地形高差，建立自外围向中心逐级递增的品质感。

设计通过连续大面积的中心景观和服务设施的分布，创造一种生活模式；梳理出高效的内部道路系统，并可伴随不同的时序灵活开发。

Client: Shum Yip Southern Land (Holdings) Co., Ltd.
Location: Dongyuan County of Heyuan
Land area: 85ha
FAR: 0.6–1.8
Building area: 514 000m²
Height: 100m
Function: Commerce, residence
International bidding: winning project

客　　户：深业南方地产（集团）有限公司
位　　置：河源市东源县
用地面积：85hm²
容 积 率：0.6~1.8
建筑面积：51.4万m²
建筑高度：100m
主要功能：商业、住宅
国际竞标：中标方案

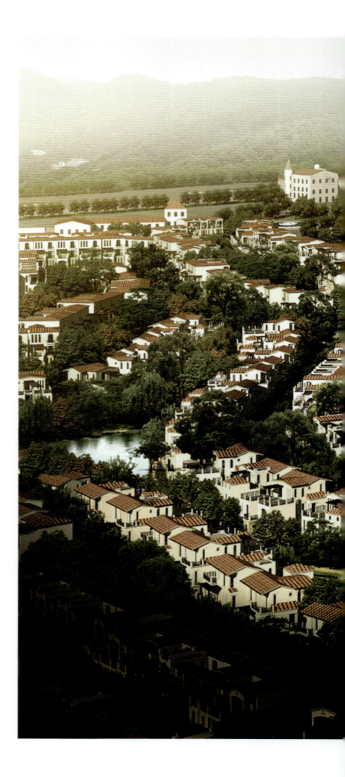

2010
LANDSCAPE
景观

01 LANDSCAPE
Design of 10 Miles Flower Valley of Yu'an-Anjing District, Yunyan, Guiyang

10 miles flower valley stretches for 7 kilometers along Nanming River with marvelous landscape and variable space.

We fully respect the original characteristics of the land and propose to create:

Biodiversity: flowers blooming and flourishing with seasons

Cultural diversity: integration of different nations and different cultures; internationalization of local culture

Spatial diversity: complex space with multilevel characteristics.

Following the principles of internationalization, sustainable development, modern ecology and local characteristic, we promote the value of the land along the river and reinforce the relation between new town and river by reasonable strategy.

In the general layout, a flexibility is reserved to future change.

贵阳云岩区十里花川公共空间及南明河两侧景观概念设计

十里花川,沿南明河总长度约7km,沿途空间启承转折,山水秀美。

整体设计依托场地原生特质,注重三个多样性营造:

生物多样性——四季轮转,十二月更替,繁花之上再生繁花;

民族多样性——多民族,多文化,地域文化的国际性道路;

空间多样性——复合空间,多层次,收放有致;

秉承国际性、持续性、现代生态化、本土个性和绿色的设计原则。

通过合理的、可持续性发展的土地开发策略提升河岸土地价值;强化门户景观空间;加强开放空间建设,山体、新城与水岸之间的联系;并关注每一个水岸空间的个性特征与丰富性。

界定空间属性和功能布局的同时,注重空间弹性,以植被空间预留未来场地属性转变的可能。

147 · LANDSCAPE

Client : Zhongtian Urban Development Group Co., Ltd.
Location : Yu'an & Anjing of Yunyan District, Guiyang
Land area : 96.7ha
Landscape area : 634 000m²
Function : Public space

客　　户：中天城投集团城市建设有限公司
位　　置：贵阳市渔安安井片区
用地面积：96.7hm²
景观面积：63.4万m²
主要功能：公共景观

02 LANDSCAPE
Design of Yuannan-Vietnam Railway Park of Kunming

Our proposal aims for creating a Yunnan-Vietnam Railway culture park integrated the characteristics of Chinese and French culture.

In proposal, we fully respect the existing landform and emphasize the balance of land value to create an organic and original landscape and an environment under sustainable development.

After taking future functional adjustment into consideration, we endow the park flexibility in space.

The shopping space in this park is also designed in a French style.

This park, with the most typical French characteristics will be a new name card of Kunming

昆明滇越铁路主题公园景观方案设计

项目设计以体现中法文化特质为核心设计理念，并以滇越铁路文化为主线。

尊重场地，强调土地综合价值平衡，通过地形处理，平衡场地挖湖后的土方平衡，最大程度降低场地土方运输量，营造有机景观，注重环境原生性，视景独特性，生态可持续性。

注重弹性空间设计，为未来公园业态调整或功能置换提供可能。

通过商业差异化定位，营造公园式购物环境，形成昆明独具特色的法式购物空间。将项目打造成昆明，云南乃至全国最具法国文化特质公园，成为昆明又一张重要城市名片。

Client : Yunnan Kunchi Real Estate Co.,Ltd
Location : Dianchi Lake Resort, Kunming
Land area : 52ha
Landscape area : 400 000m²
Function : Urban park

客　　户：云南堃驰房地产有限公司
位　　置：昆明市滇池度假区
用地面积：52hm²
景观面积：40万m²
主要功能：城市公园

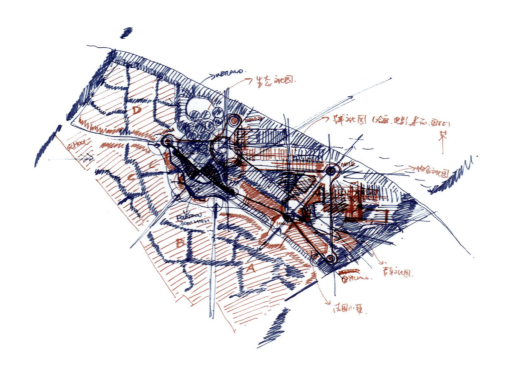

155 · LANDSCAPE

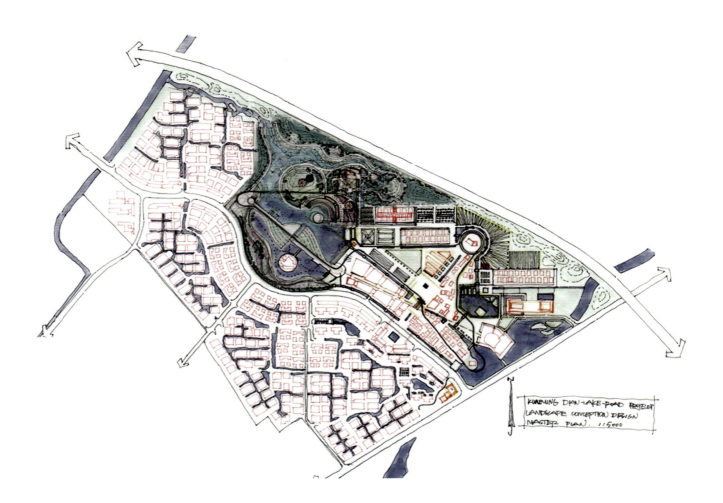

03 LANDSCAPE
Design of Green Land Park on the West of
Guiyang International Conference & Exhibition Center

The concept of "ribbon" is implanted to the pedestrian system of Green Park. It connects Guanshan Park and Guiyang International Conference& Exposition Center, as the 'nature' and 'city' are integrated together. People could ramble through the park and enjoy the beautiful scenery and quietness close to city.

贵阳国际会议展览中心西侧绿地公园景观设计

以"飘带"作为绿地公园景观步行系统的概念，连接观山公园与贵阳国际会议展览中心两种不同功能属性的公共空间，象征连接"自然"与"城市"，让人们随着步道蜿蜒穿梭于美丽自然的生态景观之中，享受最靠近繁华都市的宁静。

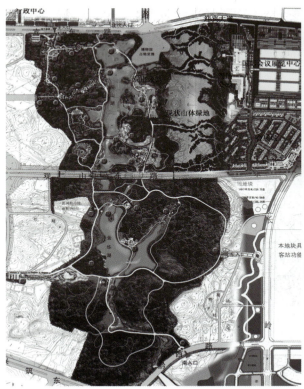
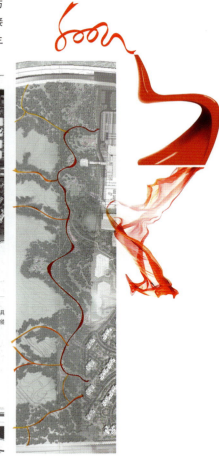

Client : Zhongtian Urban Development Group Co., Ltd.
Location : Jinyang district, Guiyang
Land area : 24ha
Landscape area : 240 000m²
Function : Public space

客　　户：中天城投集团股份有限公司
位　　置：贵阳市金阳新区
用地面积：24hm²
景观面积：24万m²
主要功能：观景休闲

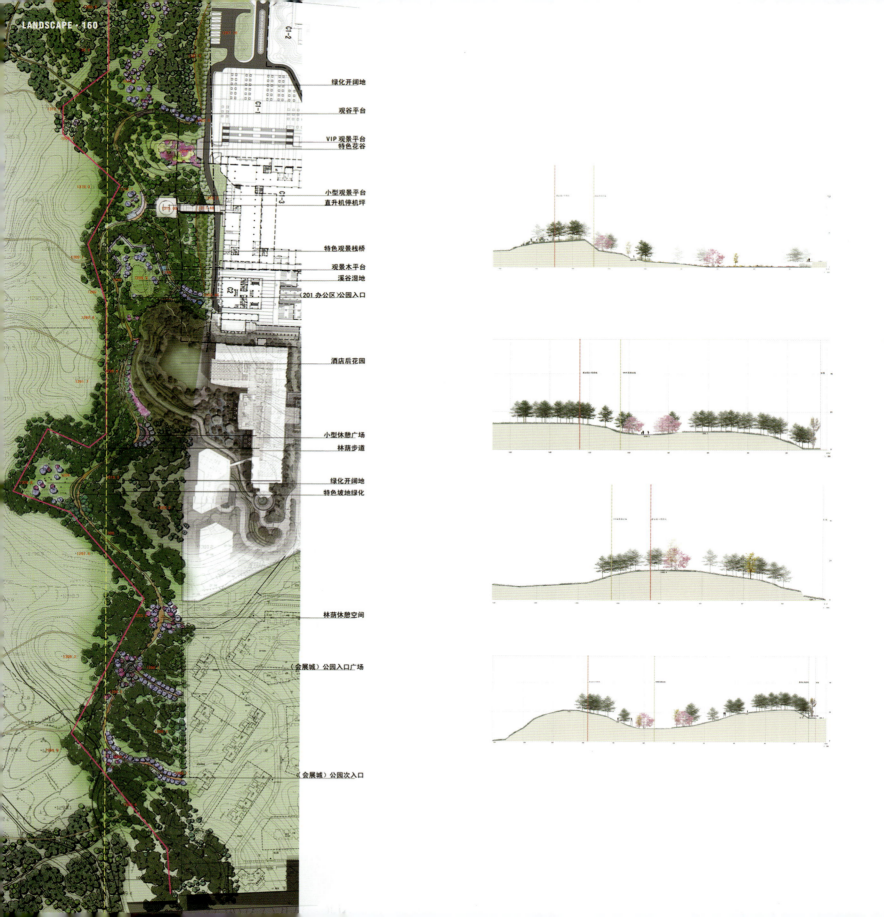

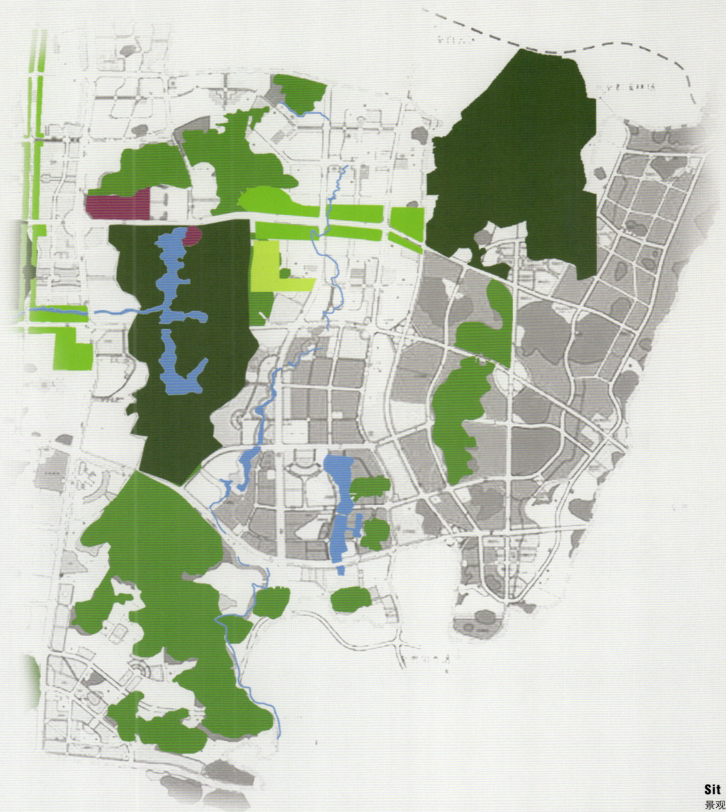

Sit plan
景观区位图

04 LANDSCAPE

Design of Konka R&D Tower of Shenzhen

Our inspiration derived from that the mainboard and chips form the nucleus of the High-Tech equipments. In the design, the plaza is the mainboard with two themes, "Technological Core" and "Window of Technology". It shines brilliantly in the lamplight during the evening. The flower beds lying on the south of the plaza look like the chips on mainboard. Palm trees are planted to give shadow and show of the magnificence of the tower. Different lamps stand on the north of the plaza to embody the sense of hign-tech and enrich the space of plaza.

深圳康佳研发大厦景观设计

高科技设备最核心的部分来自于小而精巧的芯片和主板，此方案的设计灵感由此而来，并且对其抽象运用，并赋予功能。整个广场仿佛一个主板，广场的中心景观，以"科技核心"、"科技之窗"两个不同的概念为主体，夜晚在灯光的映射下更加晶莹剔透。广场南部是整齐的花池，宛如主板上成组的芯片；花池中种植的挺拔棕榈树，在为人们带来绿荫的同时衬托出主楼的气势。广场以北阵列式的灯具小品，丰富广场上的空间，体现高科技的涵义。

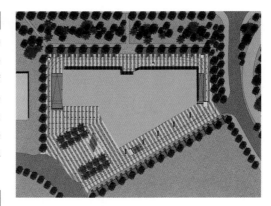

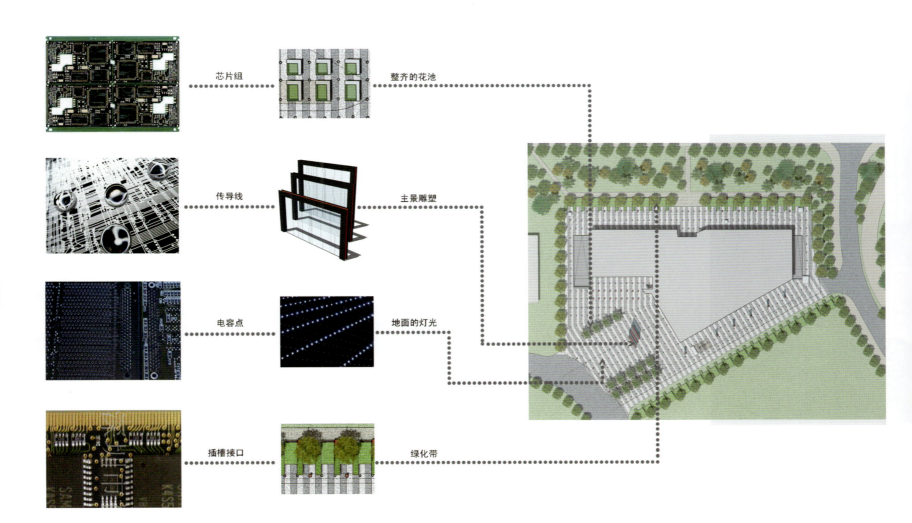

165 · LANDSCAPE

Client : Konka Group Co., Ltd
Location : Nanshan district, Shenzhen
Land area : 0.96ha
Landscape area : 11 200m²
Function : Office
Under construction : Under construction

客　　户：康佳集团股份有限公司
位　　置：深圳市南山区
用地面积：0.96hm²
景观面积：1.12万m²
主要功能：办公
在建项目：在建中

05 LANDSCAPE
Design of Finance Tower of Changfu of Shenzhen

According to the golden ratio, our architects endeavored to create a tall and erect architectural form for the tower. As the glass is the main material for the façade, the tower looks like "the waterfall pouring down from heaven". To be integrated with the architecture, we proposed the concept of "green waterfall" in landscape design. Taking full advantage of the elevation difference between the site and city roads, we utilized stepped green land to create the image that the water is spreading out after fall. It integrated also the existing public green lands lying in the north and south of the tower to form a unified landscape.

深圳长富金茂大厦景观设计

长富金茂大厦的建筑设计力图创造修长挺拔的建筑体态，依据黄金分割比塑造出更为高耸流动的上升动态，以玻璃为主的建筑群流光溢彩。远观与近赏，恰似一抹飞瀑倾空而泻，大有"飞流直下三千尺"之豪迈。在景观设计方面，借鉴和沿用了建筑设计的概念，提出"绿瀑"理念，整个场地的设计利用建筑场地与城市道路之间的高差，以台阶式绿地进行处理，仿佛瀑布流下后向四周扩散，由广场跌向周边。为与基地南北两侧已建成公共绿地衔接过渡更为自然，广场空间上采用矩阵式的绿化形势，视觉上将南北绿地连成整体，配合台阶式草坪的过渡，仿若将绿色扩散到周边，与设计主题紧密相扣。

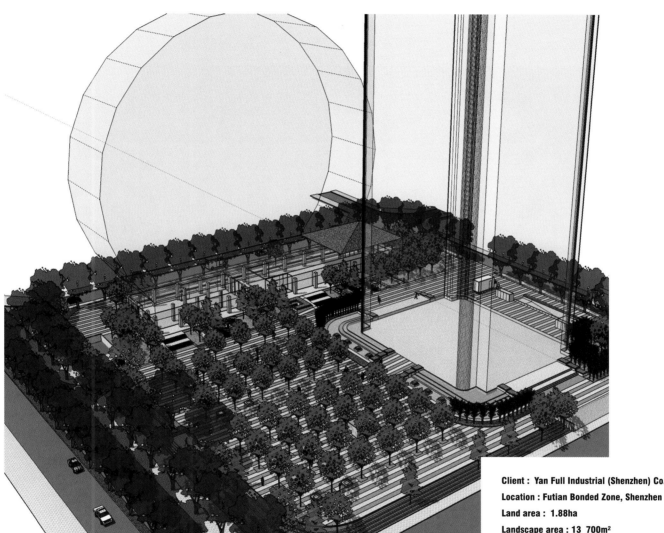

Client : Yan Full Industrial (Shenzhen) Co., Ltd
Location : Futian Bonded Zone, Shenzhen
Land area : 1.88ha
Landscape area : 13 700m²
Function : Commerce, office
Under construction : Under construction

客　　户：杨富实业(深圳)有限公司
位　　置：深圳市福田保税区
用地面积：1.88hm²
景观面积：1.37万m²
主要功能：商务、办公
在建项目：在建中

169 · LANDSCAPE

06 LANDSCAPE
Design of the Entry and Retail Street of Shum-yip New Shoreline Community of Shenzhen

Breathing Land

As people attach more and more attention to the quality of environment along with the development of society and the improvement of living conditions, the landscape design based on environment protection as well as ecological and durable development is becoming the fist choice of people and designers. In this reconstruction project, followed by principle of 'ecology', we create a green and comfortable shopping space for people.

深圳深业新岸线入口及商业街景观改造设计

呼吸的土地：

随着社会的进步和生活水平的提高，人们对环境品质的要求更高，环保的、生态的、可持续发展的景观越来越受人们的亲睐，成为当代景观设计的趋势，本案为住宅区商业街改造工程，以休闲、绿色、生态为设计理念，力求为住区居民及购物者创造一种绿色、休闲的景观及购物空间。

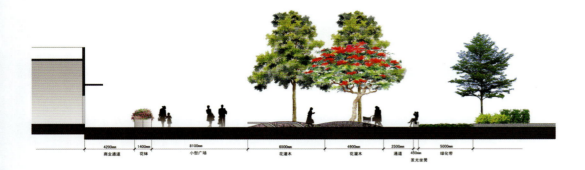

Client : Shum Yip Southern Land (Holdings) Co.,Ltd.
Location : Bao'an District, Shenzhen
Land area : 1.6ha
Landscape area : 15 800m²
Function : Retail street and plaza

客　　户：深业南方（集团）房地产开发有限公司
位　　置：深圳市宝安区
用地面积：1.6hm²
景观面积：1.58万m²
主要功能：商业街及商业广场

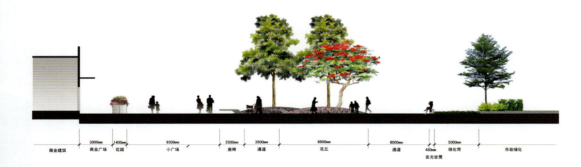

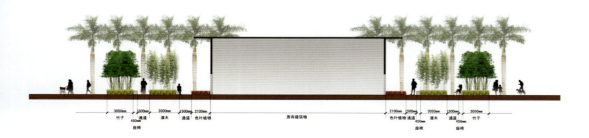

173 · LANDSCAPE

07 LANDSCAPE
Design of Blocks 6&7# of Dream Land of Guiyang

Following the general planning and architectural conception of New Century City, we create a more natural and ecological modern living environment for Blocks 6&7# in where people could 'reside in forest, enjoy in forest and breathe in forest'.

We respect the existing landform and recover the land with different plantations. We reduce the unfavorable impacts of elevation on functions by different technical measures. On the other side, we create a visual corridor and a typical montane landscape by taking advantage of elevation. We analyzed and deconstructed the architecture design to realize integration between building and landscape.

贵阳中天世纪新城6、7组团景观设计

延续规划建筑"居家山林、感受山林、呼吸山林"的总体设计理念，呼应并依托五组团商业"中心之眼"、"呼吸之眼"的功能及定位，将6、7组团打造成更为自然生态的居住组团，从而体现"居家山林、感受山林和呼吸山林"的自然、现代的山地住宅景观。尊重场地地形关系，还原场地丰富的植被环境，通过技术手段弱化场地高差对使用功能的影响，并借助地形高差形成良好的山地景观空间及视觉通廊。分析、解构建筑的设计语言，应用于景观空间布局及构筑物、小品等细节设计中，达到建筑-景观风格的协调统一。

Client： Zhongtian Urban Development Group Co., Ltd.
Location： Nanming District, Guiyang
Land area： 10.4ha
Landscape area： 48 800m²
Function： Landscape of residence

客　　户：中天城投集团股份有限公司
位　　置：贵阳市南明区
用地面积：10.4hm²
景观面积：4.88万m²
主要功能：住区景观

08 LANDSCAPE
Design for the Kindergarden of Block 4 of Dream Land of Guiyang

Paradise of children

What's the impact of colors on children? As according to the relevant research result, good-looking colors (such as blue, yellow, olivine, orange, purple, red and etc) have positive influences on children's IQ, we added some colors to the building by landscape design. At the same time, we created colorful space with flower patterns and a graffiti wall for children. We hope this kindergarten will be a real paradise for children.

贵阳中天世纪新城4组团幼儿园景观设计

孩子的欢乐花园：

色彩对儿童的影响有多少？本案幼儿园景观设计在建筑素雅颜色的基调上，增加了"好看"的颜色（如：蓝、黄、黄绿、橙、紫、红，根据专业机构研究成果，此"好看"的颜色对提高孩子们的智商有积极的影响）。同时，解构花朵这一原形，设计色彩丰富的花朵图案空间；利用高差，设计"几米漫画"的涂鸦墙，营造色彩缤纷的孩子活动场所，成为孩子梦想的欢乐花园。

Client : Zhongtian Urban Development Group Co., Ltd.
Location : Nanming District, Guiyang
Land area : 0.44ha
Landscape area : 4 400m²
Function : Landscape of kindergarden

客　　户：中天城投集团股份有限公司
位　　置：贵阳市南明区
用地面积：0.44hm²
景观面积：0.44万m²
主要功能：幼儿园景观

2010作品附录
2010 PROJECTS

索引

184 规划作品附录

186 建筑作品附录

188 景观作品附录

Index

184 **Appendix of planning works**

186 **Appendix of architectural works**

188 **Appendix of landscape works**

规划作品附录
Planning

* 已竣工或建造中项目
* Projects constructed or under construction

012

* 贵阳云岩区渔安安井片区城市设计
2010年，委托设计，用地面积755 hm²，
建筑面积650万m²
Planning and Architecture Design of Yu'an & An Jing of Yunyan District in Guiyang
2010 Mandated project Land area: 755 ha
Building area: 6 500 000 m²

018

武汉新区四新生态新城"方岛"区域城市设计
2010年，国际竞标第二名，用地面积268 hm²
建筑面积400万m²
Urban Design of "Square Island" of Sixin Ecological Town in the New Area of Wuhan
2010 Second place of international bidding
Land area: 268 ha Building area: 4 000 000 m²

022

台湾高雄海洋文化及流行音乐中心新建工程国际赛
2010年，国际竞标，用地面积10 hm²，建筑面积7.1万m²
Architecture Design of Kaohsiung Maritime Culture & Popular Music Center
2010 International bidding Land area: 10 ha
Building area: 71 000 m²

028
*

佛山西站综合交通枢纽概念性规划建筑设计
2010年，国际竞标中标方案，用地面积约31 hm²
建筑面积约8万m²
Planning and Architecture Design of Foshan West Station
2010 Winning project of international bidding
Land area: 31 ha Building area: 80 000 m²

034

长春南部新城净月西区生态商务金融中心（EBD）城市设计
2010年，国际竞标第二名，用地面积319.5 hm²，
建筑面积574.5万m²
Urban Design of EBD of Jingyue District, Changchun
2010 Second place of international bidding
Land area: 319.5 ha Building area: 5 745 000 m²

036

花溪山水度假旅游城贵阳花溪概念性城市设计
2010年，委托设计，用地面积108.5 km²，
建筑面积5800万m²
Conceptual Urban Design of Huaxi Tourism Resort Guiayang
2010 Mandated project Land area: 108.5 km²
Building area: 58 000 000 m²

040

中粮地产（集团）深圳宝安61区概念性规划建筑设计
2010年，国际竞标，用地面积 2.4 hm²，
建筑面积9.8万m²
Conceptual Planning and Architecture Design of Zone 61 of COFCO Real Estate, Bao'an District, Shenzhen
2010 International bidding Land area: 2.4 ha
Building area: 98 000 m²

044

贵阳星云家电城项目城市设计及单体建筑概念性方案设计
2010年，委托设计，用地面积 3.25 hm²
建筑面积52.7万m²
Conceptual Urban Planning & Architecture Design of Xingyun Home Appliances City of Guiyang
2010 Mandated project Land area: 3.25 ha
Building area: 527 000 m²

050

航天成都城上城规划、建筑设计
2010年，国际竞标中标方案，用地面积4.27hm²，
建筑面积30万m²，高度70m
Planning and Architecture Design of Aerospace Town in Chengdu
2010 Winning project of international bidding Land area: 4.27 ha
Building area: 300 000 m² Height: 70 m

056　　　　　　　　　060

圳半岛城邦三四五期详细蓝图修编
10年，委托设计，用地面积 26.42 hm²，建筑面积93万m²，
筑高度100~200m
tailed Blueprint Modification of Peninsula Residential
mmunity in Shenzhen
10 Mandated project Land area:26.42 ha
ilding area: 930 000 m² Height: 100-200 m

* 福州中央商务中心安置房规划建筑设计
2010年，中标方案，用地面积 4.75 hm²，
建筑面积23万m²，建筑高度<100m
Planning and Architecture Design of Resettlement
Housing of Fuzhuo CBD
2010 Winning project Land area: 4.75 ha
Building area: 230 000 m² Height: <100 m

* 贵阳市河滨剧场片区旧城改造项目
2010年，委托设计，用地面积13hm²，
建筑面积100万m²
Urban Renewal of Waterfront Theatre Area of Guiyang
2010 Mandated project Land area: 13ha
Building area: 1 000 000m²

* 贵阳市人民剧场片区旧城改造项目
2010年，委托设计，用地面积5.7hm²，
建筑面积60万m²
Urban Renewal of The People Theatre Area of Guiyang
2010 Mandated project Land area: 5.7ha
Building area: 600 000m²

家龙工业区04-05-17地块改造设计
10年，委托设计，用地面积1.5hm²，
筑面积7.9万m²
ban & Archiectural Design for the Reconstruction of
ot 04-05-17 of Majialong Industrial Zone,Shenzhen
10 Mandated project Land area: 1.5ha
ilding area: 79 000m²

* 贵阳渔安安井片区原回迁区D组团地块修建性详细
规划
2010年，委托设计，用地面积9.2hm²，
建筑面积22万m²
Constructive Detailed Planning of Bloc D of Anjing,
Yuan District, Guiyang
2010 Mandated project Land area: 9.2ha
Building area: 220 000m²

建筑作品附录
Architecture

* 已竣工或建造中项目
* Projects constructed or under construction

066

珠海港珠澳大桥·香港口岸国际概念设计竞赛
2010年，国际竞标，用地面积 10 hm²，
建筑面积40万m²
HongKong-Zhuhai-Macao Bridge · Hongkong Boundary Crossing Facilities International Design Ideas Competition
2010 International bidding Land area:10 ha
Building area: 400 000 m²

076

深圳市南山文化（美术）馆方案设计
2010年，竞标方案，用地面积 0.61 hm²，
建筑面积2.8万m²，高度40m
Architecture Design of Nanshan Culture(Arts) Museum of Shenzhen
2010 Bidding project Land area: 0.61 ha
Building area: 28 000 m² Height: 40 m

082

深圳创维"半导体设计中心"建筑方案设计
2010年，国际竞标第二名，用地面积 1.7 hm²，
建筑面积12.6万m²，高度100m
Architecture Design of Skyworth Semiconductor Design Center of Shenzhen
2010 Second place of international bidding
Land area: 1.7 ha Building area: 126 000m²
Height: 100 m

090

深圳市高新技术企业联合总部大厦方案设计
2010年，设计竞标第二名，用地面积 0.5 hm²，
建筑面积8.8万m²，高度100m
Architecture Design of United Headquarter of High-Tech Enterprises of Shenzhen
2010 Second place of international bidding
Land area: 0.5 ha Building area: 88 000 m²
Height: 100 m

094

深圳地铁一号线深大站综合体上盖物业方案设计
2010年，国际竞标，用地面积 0.98 hm²，
建筑面积9.8万m²，高度200m
Architecture Design of Building Complex above Shenzhen University Station of Metro Line No.1
2010 International bidding Land area: 0.98 ha
Building area: 98 000 m² Height: 200 m

096

深圳移动生产调度中心大厦概念性方案设计
2010年，国际竞标，用地面积 0.66 hm²，
建筑面积1.93万m²，高度50m
Architecture Design of Shenzhen Mobile Tower
2010 International bidding Land area: 0.66 ha
Building area: 19 300 m² Height: 50 m

102

深圳罗湖档案管理中心建筑设计
2010年，设计竞标第二名，用地面积 0.66hm²，
建筑面积3.5万m²，高度60m
Architecture Design of Luohu Archives Center Shenzhen
2010 Second place of international bidding
Land area: 0.66ha Building area: 35 000m²
Height: 60 m

108

深圳实验学校中学部扩建工程方案设计
2010年，竞标方案，用地面积 0.70 hm²，
建筑面积1.84万m²，高度46m
Architecture Design of Expansion Project of Shenzhen Experimental Middle School
2010 Bidding project Land area: 0.70 ha
Building area: 18 400 m² Height: 46 m

116

深圳创业投资大厦建筑设计
2010年，竞标方案，用地面积 0.52 hm²，
建筑面积8.06万m²,高度150m
Architecture Design of VC & PE Tower of Shenzhen
2010 Bidding project Land area: 0.52 ha
Building area: 80 600 m² Height: 150 m

118

深圳市青少年活动中心改扩建方案设计
2010年，国际竞标，用地面积 2.8 hm²，
建筑面积6.3万m²，高度50m
Architecture Design of Shenzhen Adolescent Activities Center
2010 International bidding Land area: 2.8 ha
Building area: 63 000 m² Height: 50 m

124

深圳市鼎和国际大厦方案设计
2010年，竞标方案，用地面积 0.82 hm²，
建筑面积10.6万m²，高度200m
Architecture Design of Shenzhen Dinghe International Tower
2010 Bidding project Land area: 0.82 ha
Building area: 106 000m² Height: 200 m

128

深圳南山建工村保障性住房建设规划建筑设计
2010年，竞标方案，用地面积22.42 hm²，
建筑面积47万m²，高度60m
Architecture Design of Resettlement Housing of Jiangong Village, Nanshan District, Shenzhen
2010 Bidding project Land area: 22.42 ha
Building area: 470 000 m² Height: 60 m

132

* 深圳南海中学建筑方案设计
2010年，中标方案，用地面积 2.4公顷，
建筑面积1.7万m²，高度24m
Architecture Design of Nanhai Middle School
2010 Winning project Land area: 2.4 ha
Building area: 17 000 m² Height: 24 m

136

* 河源深业东江波尔多皇家庄园规划建筑设计
2010年，国际竞标中标方案，用地面积 85 hm²，
建筑面积51.4万m²，高度100m
Conceptual Planning of Shum Yip·Bordeaux Royal Manor（Living Service Area of Dongyuan County of Heyuan, Guangdong）
2010 Winning project of international bidding Land area: 85ha Building area: 514 000 m² Height: 100 m

* 贵阳市城乡规划展览馆建筑设计
2010年，委托设计，用地面积3.16hm²，
建筑面积20万m²
Architectural Design of Guiyang Urban & Rural Planning Exhibition Center
2010 Mandated project Land area: 3.16ha
Building area: 200 000 m²

* 鹏基半山名苑1期北区别墅立面施工图修改
2010年，委托设计，用地面积6hm²，
建筑面积2.3万m²
Construction Drawings Modification for the Villa Façade of Pengji Hillside Villas (Phase 1)
2010 Mandated project Land area: 6ha
Building area: 23 000 m²

鹏瑞中心·深圳湾1号建筑设计
2010年，委托设计，用地面积4.6hm²，
建筑面积27万m²
Architecture Design of Pan One Center-Shenzhen Bay NO.1
2010 Mandated project Land area: 4.6ha
Building area: 270 000 m²

辽宁广电中心及北方传媒文化产业园建筑方案设计
2010年，用地面积20hm²，
建筑面积50万m²
Architectural Design of Liaoning Broadcasting Center & Northern Media Culture Park
2010 Land area: 20ha
Building area: 500 000m²

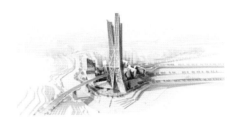
* 贵阳十里花川概念性建筑设计项目研究一
2010年，内部研究
Conceptual Design of Guiyang "Ten Miles Flower Valley"
2010 Research project

景观作品附录
Landscape

* 已竣工或建造中项目
* Projects constructed or under construction

146

* 贵阳云岩区十里花川公共空间及南明河两侧景观概念设计
2010年，委托设计，用地面积 96.7 hm²，
景观面积63.4万m²
Landscape Design of 10 Miles Flower Valley of Yu'an-Anjing District, Yunyan, Guiyang
2010 Mandated project Land area: 96.7 ha
Landscape area: 634 000 m²

152

* 昆明滇越铁路主题公园景观方案设计
2010年，委托设计，用地面积 52 hm²，
景观面积40万m²
Landscape Design of French New Station Park of Dian Lake of Kunming
2010 Mandated project Land area: 100ha
Landscape area: 400,000 m²

158

* 贵阳国际会议展览中心西侧绿地公园景观设计
2010年，委托设计，用地面积 24 hm²，
景观面积24万m²
Landscape Design of Green Land Park on the west Guiyang International Conference & Exhibition Center
2010 Mandated project Land area: 24 ha
Landscape area: 240,000 m²

162

* 深圳康佳研发大厦景观设计
2010年，委托设计，用地面积 0.96 hm²
景观面积1.12万m²
Landscape Design of Konka R&D Tower of Shenzhen
2010 Mandated project Land area: 0.96 ha
Landscape area: 11 200 m²

166

* 深圳长富金茂大厦景观设计
2010年，委托设计，用地面积 1.88 hm²，
景观面积1.37万m²
Landscape Design of Finance Tower of Changfu of Shenzhen
2010 Mandated project Land area: 1.88 ha
Landscape area: 13700 m²

170

* 深圳深业新岸线入口及商业街景观改造设计
2010年，委托设计，用地面积 1.6 hm²，
景观面积1.58万m²
Landscape Design of the Entry and Retail Street Shum-yip New Shoreline Community of Shenzhen
2010 Mandated project Land area: 1.6 ha
Building area: 15 800m²

174

* 贵阳中天世纪新城6、7组团景观设计
2010年，委托设计，用地面积 10.4 hm²，
景观面积4.88万m²
Landscape Design of Blocks 6&7# of New Century City of Guiyang
2010 Mandated project Land area: 10.4 ha
Landscape area: 48 800 m²

180
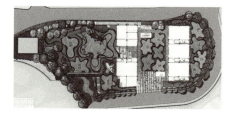
* 贵阳中天世纪新城4组团幼儿园景观设计
2010年，委托设计，用地面积 0.44 hm²，
景观面积0.44万m²
Landscape Design for the Kindergarden of Block 4 of New Century City of Guiyang
2010 Mandated project Land area: 0.44 ha
Landscape area: 4400 m²

184
* 福州市闽江北岸中央商务中心城市广场景观设计
2010年，竞标方案，景观面积5.6万m²
Landscape Design of City Plaza of CBD Northern Bank of Minjiang River, Fuzhou
2010 Bidding project
Landscape area: 56 000 m²

往年作品附录
ANNEX

索引

192 规划

200 公建

207 住宅

212 景观

Index

192 **Planning**

200 **Public Building**

207 **Residence**

212 **Landscape**

规划
Planning

* 已竣工或建造中项目
* Projects constructed or under construction

* 贵阳国际会议展览中心修建性详细规划
2009年，委托设计，用地面积51.8hm²，
建筑面积88.88万m²
Construction planning of Guiyang International
Conference & Exposition Center
2009 Mandated project Land area: 51.8 ha
Building area: 888800m²

* 中山金源花园规划设计
2009年，中标方案，用地面积21.22hm²，
建筑面积79.26万m²
Planning & Architectural Design of
Jin Yuan Garden Residence in Zhongshan
2009 Winning project Land area: 21.22ha
Building area: 792600m²

南方科技大学和深圳大学新校区拆迁安置项目
-商住综合区规划设计
2009年，竞标方案，用地面积17.77hm²，建筑面积60万m²
Removing and Resettlement of New Campus of South
University of Science and Technology and Shenzhen
University - Planning & Architectural Design for Commerce
and Residence
2009 Bidding project. Land area: 17.77ha
Building area: 600000m²

河源万绿湖风景区首期规划设计
2009年，委托设计，用地面积32hm²
Planning Design of Wanlv Lake Holiday
and Tourism Scenic Area (Phase II), Heyuan
2009 Mandated project Land area: 32ha

安徽铜陵一号地块概念规划
2009年，委托设计，用地面积27.7hm²，
建筑面积25万m²
Conceptional Planning of 1# Auction
Site of Tongling, Anhui
2009 Mandated project Land area: 27.7ha
Building area: 250000m²

深圳光明新区文化艺术中心及光明医院规划研究
2009年，委托设计，用地面积29.6hm²
Planning Research for the Sites of Center
and Hospital of Guangming New District
2009 Mandated project Land area: 29.6ha

深圳远致创业园规划设计
2009年，竞标方案，用地面积12.17hm²，
建筑面积60.67万m²
Planning & Architectural Design of Yuanzhi
Innovation Park
2009 Bidding project Land area: 12.17ha
Building area: 606,700m²

深圳观澜横坑招拍卖地块前期概念规划
2009年，委托设计，用地面积15.78hm²，
建筑面积28.57万m²
Conceptional Planning of the Auction Site
in Hengkeng of Guanlan, Shenzhen
2009 Mandated project Land area: 15.78ha
Building area: 285700m²

贵阳云岩区渔安安井片区规划设计
2009年，委托设计，用地面积1200hm²，
建筑面积450万m²
Planning Design of Yu'an & An Jing
of District, Guiyang
2009 Mandated project Land area: 1200ha
Building area: 4500000m²

金地东莞黄江居住社区概念性规划设计
2009年，国际竞标，用地面积25hm²，
建筑面积28万m²
Planning Design of Gemdale Huangjiang
Residential Community
2009 International bidding Land area: 25ha
Building area: 280000m²

深圳光明新区招拍卖地块前期概念规划
2009年，委托设计，用地面积9.07hm²，
建筑面积18.15万m²
Conceptional Planning of the Auction Site
Guangming New District, Shenzhen
2009 Mandated project Land area: 9.07ha
Building area: 181500m²

深圳市后海中心区东滨路项目概念方案设计
2009年，委托设计，用地面积4.6hm²，
建筑面积35万m²
Conceptional Urban & Architectural
Design for Dongbin Road Project, Shenzhen
2009 Mandated project Land area: 4.6ha
Building area: 350000m²

贵阳新火车站片区及商业金融区城市设计
2009年，委托设计，用地面积877.16hm²，
建筑面积1600万m²
Urban Design of Guiyang New Railway
Station Area & Commercial and Financial Area
2009 Mandated project Land area: 877.16ha
Building area: 16000000m²

惠州仲恺高新区总部经济区规划设计
2009年，竞标方案，用地面积5.42hm²，
建筑面积17.55万m²
Planning & Architectural Design of Headquarters
Economic Zone of Huizhou Zhongkai
High-Tech Incustrial Park
2009 Bidding project Land area: 5.42ha
Building area: 175500m²

深圳机场开发区西区"天空之城"概念性规划设计
2009年，中标方案，用地面积55hm²
建筑面积130万m²
Conceptional Planning Design of Sky City
(west area of Shenzhen Airport
Development Zone), Shenzhen
2009 Winning project Land area: 55ha
Building area: 1300000m²

深圳中粮宝安61区概念性规划设计
2009年，委托设计，用地面积2.43hm²，建筑面积6.79万m²
Conceptional Planning Design of Zone 61 of
Bao'an District, Shenzhen
2009 Mandated project Land area: 2.43ha
Building area: 679000m²

深圳南澳鹅公村改造规划设计
2009年，竞标方案，用地面积10hm²
Reconstruction Planning & Conceptional
Architectural Design for E Gong Village, Shenzhen
2009 Bidding project Land area: 10ha

深圳高新园区软件产业基地规划设计（第三、四标段）
2009年，竞标方案，用地面积14.9hm²，
建筑面积61万m²
Planning & Architectural Design of Software
Industry Base (section III & IV) of Shenzhen
High-Tech Industrial Park
2009 Bidding project Land area: 14.9ha
Building area: 610000m²

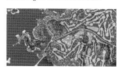
泉州洛江现代世界农业生态休闲观光园概念规划
2009年，委托设计，用地面积510.8hm²，
建筑面积76.42万m²
Concept Planning of Ecological Town
for Luojiang Modern Agricultural Tourism,
Quanzhou
2009 Mandated project Land area: 510.8ha
Building area: 764200m²

南方科技大学和深圳大学新校区拆迁安置项目-产业园区规划设计
2009年，竞标方案，用地面积15.25hm²，建筑面积63.10万m²
Removing and Resettlement of New Campus of South
University of Science and Technology and Shenzhen
University-Planning & Architectural Design
of Industry Park
2009 Bidding project Land area: 15.25ha
Building area: 631000m²

深圳市光明新区保障性住房规划设计
2008年，用地面积4.12hm²，
建筑面积13万m²
Detailed Planning of Social Community of
Guangming New Town, Shenzhen
2008 Land area: 4.12ha
Building area: 130000m²

昆明尚居五甲塘概念性总体规划
2008年，用地面积62.52hm²，
建筑面积57.34万m²
Conceptual Planning of
Wujiatang Community of S-Home, Kunming
2009 Land area: 65.52ha
Building area: 573400m²

深圳赛格日立工业区升级改造城市设计
2008年，用地面积13hm²，
Urban Design of Renewal of
Seg Hitachi Industry Zone in Shenzhen
2008 Land area: 13ha

深圳人才园规划设计
2008年，用地面积3.61hm²
建筑面积8.32万m²
Detailed Planning of
Human Resources Park, Shenzhen
2008 Land area: 3.61ha
Building area: 83200m²

珠海歌剧院规划设计
2008年，用地面积42hm²，
建筑面积4.3万m²
Urban Design of Zhuhai Opera House
2008 Land area: 427ha
Building area: 43000m²

成都怡湖玫瑰湾详细规划设计
2008年，用地面积12.57hm²，
建筑面积62.5万m²
Detailed Planning of Yihu Rose Bay
Community, Chengdu
2008 Land area: 12.57ha
Building area: 625000m²

深圳市光明新区科技公园周边地区整体城市设计暨行政中心详细城市设计
2008年，用地面积520hm²
Urban Design of Science & Technology
Park   and Administration Center of
Guangming New Town, Shenzhen
2008 Land area: 520ha

长春市高新区C-6地块住宅概念规划设计
2008年，用地面积8.57hm²，
建筑面积15.97万m²
Detailed Planning of Residential Community of
High-Tech C-6 Block, Changchun
2008 Land area: 8.57ha
Building area: 159701m²

深圳市光明新区公明文化艺术和体育中心规划设计
2008年，用地面积9.9hm²，
建筑面积5.09万m²
Urban Design of Gongming Culture, Art & Sports
Center of Guangming New Town, Shenzhen
2008 Land area: 9.9ha
Building area: 50900m²

深圳龙岗坪山街道宝山第二工业区改造概念性规划设计
2008年，用地面积74.92hm²，
建筑面积225.72万m²
Urban Design of Renewal of Baoshan 2nd Industry
Zone in Pinshan of Longgang District, Shenzhen
2008 Land area: 74.92ha
Building area: 2257200m²

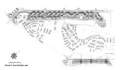
贵州遵义商业街项目概念规划设计
2008年，用地面积11.69hm²，
建筑面积60万m²
Conceptual Planning of Zunyi
Commercial Street, Guizhou
2008 Land area: 11.69ha
Building area: 600000m²

深圳天鹅堡三期概念规划设计
2008年，用地面积11.17hm²，
建筑面积20万m²
Conceptual Planning & Design of
OCT Swan Castle (Phase3), Shenzhen
2008, Land area: 11.17ha
Building area: 200000m²

宿州综合体概念性规划设计
2008年，用地面积21.3hm²，
建筑面积67.92万m²
Conceptual Planning of Suzhou Complex
2008 Land area: 21.3ha
Building area: 679200m²

长春市净月区梧桐街住宅规划设计
2008年，用地面积34.8hm²，
建筑面积45万m²
Planning Design of Residential Community of
Phoenix Tree Street of Jingyue District in Changchun
2008 Land area: 34.8ha
Building area: 450000m²

中国饮食文化城城市设计
2008年，用地面积220hm²，
建筑面积38万m²
Urban Design of Chinese Gastrologic and
Cultural Town in Shenzhen
2008 Land area: 220ha
Building area: 380000m²

深圳市南油购物公园城市设计
2008年，国际竞标第一名，
用地面积13hm²，建筑面积45.4万m²
Urban Design of
Nanyou Shopping Park, Shenzhen
2008 First prize of international bidding

深圳光明新城万丈坡居住小区详细蓝图设计
2008年，详蓝编制范围22hm²，规划研究范围49.6hm²，
建筑面积53.23万m²
Detailed Blueprint Design of Wanzhangpo Residential
Community of Guangmin New Town, Shenzhen
2008 Scope of detailed blueprint: 22ha
Scope of planning research: 49.6ha
Building area: 532,300m²

深圳南澳月亮湾海岸带景观改造设计概念规划
2008年，用地面积14.2hm²
Conceptual Plannning Design for Landscape
Reconstruction of Coastal Zones of Nan'ao, Shenzhen
2008 Land area: 14.2ha

深圳市光明新区中央公园概念规划方案
2008年，用地面积237hm²
Conceptual Planning Design of Central
Park of Guangming District, Shenzhen
2008 Land area: 237ha

珠海五洲花城二期概念性规划设计
2008年，用地面积16.5hm²，
建筑面积60万m²
Conceptual Planning of Five Continental
Residence Garden, Zhuhai
2008 Land area: 16.5ha
Building area: 600000m²

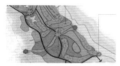

朗钜呼和浩特高尔夫社区概念规划
2008年，用地面积282hm²，
建筑面积150万m²
Conceptual Planning of
Large's Golf Residence Community, Hohhot
2008 Land area: 282ha
Building area: 1500000m²

贵阳中天集团中华北路片区概念规划设计
2008年，用地面积32.67hm²，
建筑面积232.16万m²
Conceptual Planning of
Zhonghua North Street Block of
Zhongtian Group, Guiyang
2008 Land area: 32.67ha
Building area: 2321600m²

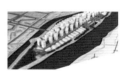

中山市东凤镇新沙岛规划设计
2008年，用地面积80.5hm²，
建筑面积80万m²
Urban Design of Xinsha Island of
Dong feng Town, Zhongshan
2008 Land area: 80.5ha
Building area: 800000m²

长沙华润含浦项目概念性规划设计
2008年，用地面积15hm²，
建筑面积45万m²
Conceptual Planning of CRL Hanpu Residential
Community, Changsha
2008 Land area: 15ha
Building area: 450000m²

成都天府华侨城二期概念规划设计
2008年，用地面积10hm²，
建筑面积15万m²
Conceptual Planning of
Tianfu OCT Residence (Phase 2), Chengdu
2008 Land area: 10ha
Building area: 150000m²

中山市坦州镇居住小区规划设计
2008年，用地面积11.4hm²，
建筑面积28万m²
Detailed Planning of Tanzhou Residential
Community, Zhongshan
2008 Land area: 11.4ha
Building area: 280000m²

* 昆明滇池旅游度假区文化公园规划设计
2007年，用地面积111hm²，
建筑面积88.8万m²
Planning Design of Cultural Park of
Dianchi Tourism & Resort Region, Kunming
2007 Land area: 111ha
Building area: 888000m²

* 惠州鹏基半山名苑规划设计
2007年，用地面积49.71hm²，
建筑面积64.1万m²，竣工日期：2009年
Planning Design of Pengji Hillside Residential
Community, Huizhou
2007 Land area: 49.71ha
Building area: 641000m² Constructed in 2009

江苏省姜堰市锦绣姜堰规划设计
2007年，用地面积29.62hm²，
建筑面积62万m²
Planning Design of Jiangyan Splendid Town,
Jiangsu
2007 Land area: 29.62ha
Building area: 620000m²

* 深圳城建·观澜居住区规划设计
2007年，用地面积16.99hm²，
建筑面积34.6万m²
Planning Design of Guanlan Residential
Community, Shenzhen
2007 Land area: 16.99ha
Building area: 346000m²

深圳中兴大梅沙培训基地规划设计
2007年，用地面积11hm²，
建筑面积11.94万m²
Planning Design of ZTE Formation
Base in Dameisha Shenzhen
2007 Land area: 11ha
Building area: 119400m²

清远新时代广场项目概念规划方案设计
2007年，用地面积55.19hm²，
建筑面积82.37万m²
Conceptual Planning Design of
New Times Plaza, Qingyuan
2007　Land area: 55.19ha
Building area: 823700m²

江苏连云港新华社区概念规划研究
2007年，用地面积97hm²，
建筑面积101.13万m²
Conceptual Planning Research of
Xinhua Residential Community, Lianyungang, Jiangsu
2007　Land area: 97ha
Building area: 1011300m²

东莞黄江伯爵山庄概念性规划设计
2007年，用地面积103.47hm²，
建筑面积36万m²
Planning Design of Earl Resort Hotel, Dongguan
2007　Land area: 103.47ha
Building area: 360000m²

深圳市观澜版画基地概念性规划
2007年，用地面积150 hm²，
建筑面积19万m²
Planning Design of Guanlan Art Print Base,
Bao'an district, Shenzhen
2007　Land area: 150ha
Building area: 190000m²

* 成都华润置地翡翠城小学规划设计
2007年，用地面积1.65m²，
建筑面积1.71万m²
Planning Design of
CRL Jade-City Primary School, Chengdu
2007　Land area: 1.65ha
Building area: 17100m²

深圳市罗湖区贝丽中学（水贝珠宝学校）规划设计
2007年，国际竞标，用地面积1.95hm²，
建筑面积2.53万m²
Planning Design of Shenzhen Beili Middle School
2007　International bidding
Land area: 1.95ha
Building area: 25300m²

* 深圳市龙岗区天安数码新城二期规划设计
2007年，国际竞标中标方案，用地面积4.63hm²，
建筑面积24.68万m²
Planning Design of Longgang Cyber Park, Shenzhen
2007　Winning project of international bidding
Land area: 4.63ha
Building area: 246800m²

佛山南海区狮山镇客运站修建性详细规划
2007年，国际竞标中标方案，用地面积10hm²，
建筑面积13.14万m²
Detailed Planning of Shishan
Long Distance Bus Station, Foshan
2007　Winning project of international bidding
Land area: 10ha
Building area: 131400m²

佛山南海区狮山镇文化体育公园规划设计
2007年，用地面积13.90hm²，
建筑面积28万m²
Planning Design of Shishan Cultural & Sport Park,
Nanshan, Foshan
2007　Land area: 13.90ha
Building area: 280000m²

珠海前山新冲路城市壹站规划设计
2007年，国际竞标，用地面积2.85hm²，
建筑面积8.31万m²
Planning Design of the First City Station
Residential Community, Zhuhai
2007　International bidding
Land area: 2.85ha
Building area: 83100m²

深圳市龙岗区"深业·坪山"居住区规划设计
2007年，国际竞标，用地面积2.83hm²，
建筑面积11.7万m²
Planning Design of 'Shum Yip Pingshan'
Residential Community, Shenzhen
2007　International bidding
Land area: 2.83ha
Building area: 117000m²

苏州太湖旅游度假区东入口区域概念性规划设计
2007年，中标方案，用地面积43hm²
Conceptual Planning of East Entrance Area of
Taihu Lake National Holiday zone, Suzhou
2007　Winning project　Land area: 43ha

* 贵阳中天世纪新城三号地块六、七组团规划设计
2007年，用地面积17.3m²，
建筑面积20万m²
Planning Design of the 6#&7# Blocks of
Zhongtian New Century City, Guiyang
200　Land area: 17.3ha
Building area: 200000m²

* 半岛城邦四、五期规划设计
2007年，用地面积11.12hm²，
建筑面积40.76万m²
Planning Design of The Peninsula (Phase 4 &5)
2007　Land area: 11.12ha
Building area: 407600m²

太古城深圳蛇口东填海区城市规划设计咨询
2007年，用地面积6.6hm²，
建筑面积23.40万m²
Planning Consultation of Grand Residential
Community in the East Filling-sea Area,
Shekou, Shenzhen
2007　Land area: 6.6ha
Building area: 234000m²

深圳中信惠州东江新城一期项目规划与建筑设计
2007年，用地面积25hm²，
建筑面积60万m²
Planning & Architecture Design of CITIC
Dongjiang New City (Phase 1), Huizhou
2007　Land area: 25ha
Building area: 600000m²

* 鹏基惠州半山名苑居住区设计
2007年，用地面积49.76hm²，
建筑面积30万m²
Planning Design of Pengji Banshan Residential
Community, Huizhou
2007　Land area: 49.76ha
Building area: 300000m²

吉林省长春市科技文化综合中心概念性规划
2006年，用地面积155.8hm²，
建筑面积67.4万m²
Conceptual Planning of Science
& Culture Center, Changchun
2006　Land area: 155.8ha
Building area: 674000m²

长沙中信新城概念规划与建筑设计
2006年，用地面积109.54hm²，
建筑面积135.47万m²
Conceptual Planning & Architecture Design
of CITIC New City, Changsha
2006　Land area: 109.54ha
Building area: 1354700m²

苏州中信太湖文化论坛规划设计
2006年，用地面积55.6hm²，
建筑面积67.5万m²
Planning Design of CITIC Taihu Lake
Cultural Forum, Suzhou
2006　Land area: 55.6ha
Building area: 675000m²

深圳迈瑞研发基地周边街区概念性城市研究
2006年，用地面积2.39hm²，
建筑面积9.49万m²
Conceptual Urban Planning of
the Surrounding Area of Mindray R&D Base
2006 Land area: 2.39ha
Building area: 94900m²

深圳CEO创意领地概念规划设计
2006年，用地面积40hm²，
建筑面积33万m²
Conceptual Planning of
Creative Experiencing Origin, Shenzhen
2006 Land area: 40ha
Building area: 330000m²

深圳市蛇口东角头片区城市设计
2006年，用地面积32.70hm²，
建筑面积54.6万m²
Urban Design of DongJiaoTou Subdistrict, Shekou, Shenzhen
2006 Land area: 32.70ha
Building area: 546000m²

东莞松山湖项目概念规划设计
2006年，用地面积9.35hm²，
建筑面积12.46万m²
Conceptual Planning of
Songshanhu Residential Community, Dongguan
2006 Land area: 9.35ha
Building area: 124600m²

* 成都天府华侨城概念性规划国际竞标
2005年，国际竞标，用地面积200hm²，
建筑面积200万m²
Conceptual Planning of Tianfu-OCT, Chengdu
2005 International bidding
Land area: 200ha
Building area: 2000000m²

深圳招商华侨城尖岗山商业中心规划设计
2005年，国际竞标，用地面积6.86hm²，
建筑面积6.5万m²
Conceptual Design of
OCT Jiangganshan Shopping Center, Shenzhen
2005 International bidding
Land area: 6.86ha
Building area: 65000m²

* 成都阳明山庄规划设计
2005年，用地面积33.34hm²，
建筑面积26.56万m²
Planning Design of Chengdu Yangming
Shanzhuang Residence
2005 Land area: 33.34ha
Building area: 265600m²

北京密云休闲度假小区概念性规划设计
2005年，用地面积8hm²，
建筑面积14.28万m²
Conceptual Planning of Beijing
Miyun Vacation Residence
2005 Land area: 8ha
Building area: 142800m²

杭州市钱江科技创业中心规划设计
2005年，用地面积11.5hm²，
建筑面积11万m²
Planning Design of
Qianjiang Technology Creative Center
2005 Land area: 11.5ha
Building area: 110000m²

* 成都天府长城二期规划设计
2005年，中标方案，用地面积6.22hm²，
建筑面积21万m²
Planning Design of
Tianfu Great Wall Residence (Phase 2), Chengdu
2005 Winning project Land area: 6.22ha
Building area: 210000m2m²

深圳市南澳海景度假中心概念性规划设计
2005年，用地面积64hm²，
建筑面积20万m²
Conceptual Planning Design of
Nan'ao Seascape Vacation Center
2005 Land area: 64ha
Building area: 200000m²

长沙苗圃项目概念规划
2005年，前期研究，用地面积80hm²
Conceptual Planning of Miaopu Project, Changsha
2005 Land area: 80ha

深圳市宝安26区旧城改造项目总体规划设计
2005年，国际竞标，用地面积23.13hm²，
建筑面积80.7万m²
General Planning for Renovation of 26# Block in Bao'an, Shenzhen
2005 International bidding,
Land area: 23.13ha
Building area: 807000m2m²

深圳蛇口东"宝能-太古城"初步规划概念设计
2005年，用地面积6.93hm²，
建筑面积18.25万m²
Conceptual Planning of
Baoneng Taigu Town, Shekou, Shenzhen
2005 Land area: 6.93ha
Building area: 182573m²

* 合肥澜溪镇A、B区建筑及规划设计
2005年，用地面积7.23hm²，
建筑面积10.63万m²
竣工日期：2007年，获奖项目
Planning Design of Section A&B of Nancy Town, Hefei
2005 Land area: 7.23ha
Building area: 106300 m²
Constructed in 2007 Awarded project

长沙天际岭项目规划建筑设计
2005年，国际竞标，用地面积38.81hm²，
建筑面积39.77万m²
Planning Design of Tianjiling Project, Changsha
2005 International bidding
Land area: 38.81ha
Building area: 397706m²

深圳市金光华"绿谷蓝溪"居住小区规划设计
2005年，用地面积19.92hm²，
建筑面积35.83万m²
Planning Design of Jingguanghua
'Lvgu Lanxi' Residential Community, Shenzhen
2005 Land area: 19.92ha
Building area: 358354m²

苏州工业园区"左岸山庭"规划与建筑概念设计
2005年，国际竞标，用地面积19.53hm²，
建筑面积39万m²
Planning Design of Hill-Yard on Left-Bank, Suzhou,
2005 International bidding
Land area: 19.53ha Building area: 390705m²

西安中海华庭居住小区规划设计
2004年，用地面积5.41hm²，
建筑面积17.3万m²
Planning Design of Zhonghai Huating Residential
Community, Xi'an
2004 Land area: 5.41ha
Building area: 173000m²

* 成都市"中海-国际社区"规划与建筑设计
2004年，用地面积132.42hm²，
建筑面积125万m²
Planning Design of
'Zhonghai-International Community', Chengdu
2004 Land area: 132.42ha
Building area: 1250000m²

* 华润置地成都翡翠城汇锦云天居住小区规划设计
2004年，国际竞标第二名，二期用地面积8hm²，
建筑面积14.56万m²，竣工日期：2007年
Detailed Planning of CRL Emerald City
Residence (Phase 2), Chengdu
2004, Second prize of international bidding,
Land area of Phase 2: 8ha
Building area: 145600m² Constructed in 2007

* 深圳市宝安西乡富通居住区概念规划设计
2004年，用地面积25.23hm²，
建筑面积50万m²
Planning Design of Futong Residential Community of
Bao'an District, Shenzhen
2004 Land area: 25.23ha
Building area: 500000m²

深圳市鸿荣源龙岗中心城规划设计
2004年，用地面积40hm²，
建筑面积60万m²
Planning Design of Hongrongyuan Longgang
Central City, Shenzhen
2004 International bidding
Land area: 40ha
Building area: 600000m²

韶关风度广场概念性规划设计
2004年，用地面积3.2hm²，
建筑面积10.25万m²
Planning Design of Fengdu Plaza, Shaoguan
2004 Land area: 3.2ha
Building area: 102500m²

北京宣武区国信大吉片规划设计
2004年，用地面积43.9hm²，
建筑面积165万m²
Planning Design of Guoxin Dajipian in Xuanwu District,
Beijing
2004 Land area: 43.9ha
Building area: 1650000m²

湖南"天健长沙芙蓉中路项目"居住区规划设计
2004年，用地面积17.12hm²，
建筑面积23.90万m²
Planning Design of Tianjian Furongzhong
Road Residence, Changsha
2004 Land area: 17.12ha
Building area: 239000m²

福州市登云山庄总体规划设计
2004年，用地面积267.4hm²，
建筑面积68.16万m²
Planning Design of Dengyun House, Fuzhou
2004 Land area: 267.4ha
Building area: 681600m²

* 深圳市宝安区26区商业公园详细蓝图设计
2004年，用地面积26.7hm²，
建筑面积65万m²，预计竣工日期：2010年
Detailed Blueprint Design of Commercial
Park of Bao'an 26# Block, Shenzhen
2004 Land area: 26.7ha
Building area: 650000m²
Construction will be finished on 2010

中央音乐学院珠海分校规划设计
2003年，国际竞标，用地面积43hm²，
建筑面积15万m²
Planning Design of National Music University-
Zhuhai Branch
2003 International bidding
Land area: 43ha Building area: 150000m²

杭州下沙开发区城市规划设计
2003年，用地面积85hm²，
建筑面积91万m²
Urban Planning of Xiasha
Development Zone, Hangzhou
2003 Land area: 85ha
Building area: 910000m²

福建晋江市行政中心规划设计
2003年，用地面积24hm²，
建筑面积8.5万m²
Planning Design of Jinjiang
Administration Center, Fujian
2003 Land area: 24ha
Building area: 85000m²

* 苏州东城郡住宅小区规划设计
2003年，用地面积6.5hm²，
建筑面积14.70万m²，竣工日期：2006年
Planning Design of Residential Community of
East County of Suzhou Industry Park
2003 Land area:6.5ha
Building area: 147000m² Constructed in 2006

深圳盐田中心区城市规划设计
2003年，城市规划设计面积18hm²，
环境景观设计面积6.50hm²
Urban Planning of Central Area of Yantian District,
Shenzhen
2003 Urban planning area: 18ha
Landscape design area: 6.50ha

* 深圳蛇口填海区大型住宅区详细蓝图规划设计
2003年，用地面积325hm²
Detailed Blueprint Design of Large Scale Residential
Community in Shekou reclamation Area, Shenzhen
2003 Land area: 325ha

珠海情侣路滨海带规划设计
2003年，国际竞标，规划环境景观面积150hm²，
海岸线6.5公里
Planning Design of Seabelt Area of
Lovers-Road, Zhuhai
2003 International bidding
Landscape design area:150ha
Coastal line : 6.5km

深圳市沙河世纪山谷住宅小区详细蓝图规划设计
2003年，用地面积18hm²，
建筑面积67万m²
Detailed Blueprint Design of Shahe Century
Valley Residential Community, Shenzhen
2003 Land area: 18ha
Building area: 670000m²

海南省三亚阳光海岸城市规划设计
2003年，国际竞标，用地面积146.67hm²
Urban Planning of Sanya Sunny-coast, Hainan
2003 International bidding
Land area: 146.67ha

深圳市南山区大冲住宅区概念规划设计
2003年，用地面积67hm²，
建筑面积132万m²
Conceptual Planning of
Dachong Residential Community, Shenzhen
2003 Land area: 67ha
Building area: 1320000m²

苏州新加坡工业园区东湖大郡二期住宅小区规划设计
2003年，用地面积10hm²，
建筑面积18万m²
Planning Design of East Lake Residence (Phase 2) of
Singapore Industry Park, Suzhou
2003 Land area: 10ha
Building area: 180000m²

深圳市华侨城集团惠州温泉度假村规划设计
2002年，用地面积700hm²
Planning Design of OCT
Spring Spa Holiday Village, Huizhou
2002 Land area: 700ha

* 福州融侨"江南水都"首期住宅小区规划设计
2002年，用地面积12.34hm²，
建筑面积18.4万m²，竣工日期：2006年
Detailed planning of Rongqiao
"Jiangnan Water Town" (Phase 1), Fuzhou
2002 Land area: 12.34ha
Building area: 184000m2 Constructed in 2006

福建招商局漳州开发区生活中心规划设计
2002年，规划面积100hm²，
首期商场用地面积4.4hm²，建筑面积4.80万m²
Planning Design of Living Center of Merchants Bureau
Zhangzhou Developing Zone, Fujian
2002 Planning area: 100ha
Shopping center site area of Phase 1: 4.4ha
Building area: 48000m²

南京将军山别墅及多层住宅小区规划设计
2002年，用地面积23.5hm²，
建筑面积16万m²
Planning Design of Mt. General Villa and
Muli-floors Residence, Nanjing
2002 Land area: 23.5ha
Building area: 160000m²

上海嘉定高尔夫社区规划设计
2002年，用地面积667hm²
Urban Design of
Jiading Golf Community, Shanghai
2002 Land area: 667ha

深圳高新技术开发区中心区规划设计
2001年，用地面积25.6hm²，
建筑面积27万m²
Planning Design of
Hi-tech Developing Area Center, Shenzhen
2001, Land area: 25.6ha
Building area: 270000m²

* 深圳市南海益田详细蓝图规划设计
2001年，国际竞标中标方案，用地面积31hm²，
建筑面积92万m²
Detailed Blueprint Design of
Nanhai Yitian Residence, Shenzhen
2001 Winning project of international bidding,
Land area: 31ha
Building area: 920000m²

温州苍南新市中心区规划设计
2001年，用地面积454hm²，
建筑面积15万m²
Planning Design of
Cangnan New City Center, Wenzhou
2001 Land area: 454ha
Building area: 150000m²

深圳国际网球中心俱乐部小区规划设计
2001年，用地面积9.8hm²，住宅建筑面积14万m²，
国际网球中心建筑面积3万m²
Planning Design of Shenzhen International
Tennis Center Club Residence
2001 Land area: 9.8ha
Building area of house: 140000m² Building
area of international tennis center: 30000m²

成都万安国际社区规划设计
2001年，概念设计，用地面积266hm²，
住宅建筑面积106万m²
Planning Design of Wan'an International
Community, Chengdu
2001, Land area:266ha
Housing area: 1060000m²

成都长城地产五洲花园住宅区规划设计
2001年，国际竞标，用地面积63hm²，
总建筑面积80万m²
Planning Design of Changcheng Five Continents
Residential Garden, Chengdu
2001 International bidding
Land area: 63ha
Building area: 800000m²

* 贵阳中天集团世纪新城联排住宅小区规划设计
2001年，建筑面积5.2万m²，竣工日期：2003年
Planning Design of Zhongtian Century
New City Townhouse, Guiyang
2001 Building area: 52000m²
Constructed in 2003

吉林市松花江两岸总体规划构思及重点地区城市设
2001年国际竞标，规划陆地面积2525hm²，
城市设计范围971hm²
General Planning of Songhuajiang
River Bank & Urban Design of Key Areas
2001 International bidding
Land area: 25.25ha
Urban design area: 971ha

济南市南部新城区城市规划设计
2001年，用地面积396hm²，
建筑面积15万m²
Urban Planning of New South District of Jinan
2001 Land area: 396ha
Building area: 150000m²

深圳罗湖百仕达山水城综合住宅小区规划设计
2001年，用地面积10hm²，住宅建筑面积28万m²，
商业建筑面积6.10万m²，酒店式公寓5万m²
Planning Design of Luohu Baishida Mountain
& Water City Residence, Shenzhen
2001 Land area:10ha Housing building area: 280000m²,
Business building area: 61000m²,
Hotel-style Apartment: 50000m²

广州新世界地产禺东西路规划设计
2001年，建筑面积14万m²
Planning Design of Yudong Xi Road Project of
New World Real Estate, Guangzhou
2001 Building area: 140000m²

深圳宝安中心城市规划设计
2000年，国际竞标，用地面积100hm²，
建筑面积8300m²，绿化面积30万m²
Urban Planning of Central Area of Bao'an
District, Shenzhen
2000 International bidding
Land area: 100ha Building area: 8300m²
Green area: 300000m²

东莞御花园住宅小区规划设计
2000年，建筑面积50万m²
Planning Design of Imperial
Garden Residence, Dongguan
2000 Building area: 500000m²

深圳市盐田区大梅沙梅西谷别墅区可行性研究
2000年，用地面积14hm²
Feasibility research for Meixi Valley Villa Area
of Dameisha, Shenzhen
2000 Land area: 14ha

北京万科青青家园住宅小区规划设计
2000年，用地面积23hm²，
建筑面积28万m²
Planning Design of Vanke Qingqing House, Beijing
2000 Land area: 23ha
Building area:280000m²

湖州市仁皇山新区城市规划设计
2000年，国际竞标，用地面积400hm²
Urban Planning of Renhuangshan New District,
Huzhou
2000 International bidding
Land area: 400ha

南海市海八路以北规划设计
2000年，用地面积174hm²，
建筑面积23.90万m²
Planning Design of the North of haiba Road, Nanhai
2000 Land area:174ha
Building area: 239044 m²

深圳盐田碧海名峰（现名天琴湾）别墅区规划设计
1999年，用地面积29hm²，建筑面积3.8万m²，
预计竣工日期：2010年
Urban Design of
Yantian Bihaimingfeng (Lyra Bay), Shenzhen
1999 Land area: 29ha Building area: 38000 m²
Construction will be finished in 2010

广西桂林市水系规划设计
1999年，国际竞标，300公顷核心范围和
600hm²影响范围
Planning Design of
Guilin Water System, Guangxi
1999 International bidding
Core area:300ha Influenced area: 600ha

* 深圳盐田菠萝山海水温泉休闲区规划设计
1999年，用地面积24hm²，
建筑面积8万m²，竣工日期：2005年
Planning Design of Yantian Pineapple Mountain
Spa Leisure Area, Shenzhen
1999, Land area: 24ha
Building area: 80000m² Constructed in 2005

杭州市钱塘江两岸新中心区规划设计
1999年，国际竞标，用地面积300hm²
Planning Design of New CBD along the
Qiantangjiang River, Hangzhou
1999 International bidding
Land area: 300ha

* 珠海珠澳海关地区规划设计
1998年，规划用地面积60hm²，
建筑面积7万m²，竣工日期：1998年
Planning Design of Zhu'ao Customs Area, Zhuhai
1998 Planning Land area: 60ha
Building area: 70000m² Constructed in 1998

珠海二横琴湾规划设计
1997年，用地面积100hm²
Planning Design of Second Hengqin Bay, Zhuhai
1997 Land area: 100ha

广西壮族自治区来宾法国电力公司厂区规划设计
1997年，建筑面积3万m²
Planning Design of Laibin French
Power Company Workshop, Guangxi
1997 Building area: 30000m²

江苏东海市圣戈班玻璃工厂和生活基地
法国专家村规划设计
1997年，用地面积18hm²，建筑面积2.5万m²
Planning Design of St. Gorban Glass Manufacturer
and Living Area French Expert Village, Donghai,
Jiangsu
1997 Land area: 18ha
Building area: 25000m²

江苏省无锡法国达能矿泉水厂和生活基地规划设计
1997年，建筑面积3万m²
Planning Design of French Daneng Mineral Water
Factory and Living Area, Wuxi, Jiangsu
1997 Building area: 30000m²

江苏省南京市师范大学校园区规划设计
1997年，国际竞标二等奖，
建筑面积25万m²
Planning Design of Nanjing Normal University
Campus, Jiangsu
1997 Second prize of international bidding
Building area: 250000m²

珠海中国国际青少年活动中心规划设计
1997年，用地面积98hm²
Planning Design of
China International Youth Center, Zhuhai
1997 Land area: 98ha

* 深圳市华侨城中西部城市综合区设计
1996年，用地面积85hm²，
建筑面积185万m²
Urban Planning of
OCT Center-West Area, Shenzhen
1996 Land area: 85ha
Building area: 1850000m²

深圳市福田区新市中心和市民中心规划设计
1996年，国际竞标，用地面积200hm²，
市政厅建筑面积10万m²
Planning Design of Futian New City
Center and Citizen Center
1996 International bidding Land area: 200ha
Building area of City Hall: 100000m²

中山市"一河两岸"规划设计
1996年，用地面积100hm²
Planning Design of
Both Shores of One River, Zhongshan
1996 Land area: 100ha

北京国门广场住宅区规划设计
1995年，建筑面积28万m²
Planning Design of National Gate
Square Residence, Beijing
1995 Building area: 280000m²

上海市浦东六里现代化生活居住区园区规划设计
1995年，建筑面积30万m²
Planning Design of Pudong Six Li Modern
Life Residence Community, Shanghai
1995 Building area: 300000m²

珠海横琴行政中心规划设计
1994年，用地面积250hm²
Planning Design of
Hengqin Administration Center, Zhuhai
1994 Land area: 250ha

海南省海口市新城市中心规划设计
1994年，国际竞标中标方案，
用地面积200hm²，建筑面积350万m²
Planning Design of
New City Center of Haikou, Hainan
1994 Winning project of international bidding
Land area: 200ha Building area: 3500000m²

* 珠海大学校园规划设计
1993年，国际竞标一等奖，用地面积176hm²，
建筑面积36万m²，竣工日期：2000年
Planning Design of Zhuhai University
1993 First Prize of International bidding
Land area: 176ha
Building area: 360000m² Constructed in 2000

公建
Public building

* 已竣工或建造中项目
* Projects constructed or under construction

* 贵阳国际会议展览中心建筑工程设计
2009年，委托设计，用地面积51.8hm²，
建筑面积88.88万m²
Architectural Design of Guiyang International
Conference & Exposition Center
2009 Mandated project Land area: 51.8 ha
Building area: 888800m²

深圳国际能源大厦建筑设计
2009年，国际竞标，用地面积0.64hm²，
建筑面积11.7万m²
Architecture Design of Shenzhen International
Energy Mansion
2009 International bidding Land area:0.64ha
Building area: 117000m²

惠州仲恺高新区总部经济区建筑设计
2009年，竞标方案，用地面积5.42hm²，
建筑面积17.55万m²
Planning & Architectural Design of Headquarters
Economic Zone of Huizhou Zhongkai
High-Tech Industrial Park
2009 Bidding project Land area: 5.42ha
Building area: 175500m²

深圳文学艺术中心建筑设计
2009年，竞标方案第二名，用地面积1.0hm²，
建筑面积6.20万m²，高度100m
Architectural Design of Shenzhen Literary
Arts Center
2009 Second prize of bidding
Land area: 1.0ha
Building area: 62000m² Height: 100m

深圳移动生产调度中心大厦概念性方案设计
2009年，竞标方案，用地面积0.56hm²，
建筑面积10万m²，高度159.6m
Architectural Design of Production Scheduling
Center Building of China Mobile (Shenzhen)
2009 Bidding project Land area: 0.56ha
Building area: 100000m² Height:159.6m

深圳远致创业园建筑设计
2009年，竞标方案，用地面积12.17hm²，
建筑面积60.67万m²，高度245m
Planning & Architectural Design of Yuanzhi
Innovation Park, Shenzhen
2009 Bidding project Land area: 12.17ha
Building area: 606700m² Height: 245m

深圳高新园区软件产业基地建筑设计（第三、四标段）
2009年，竞标方案，用地面积14.9hm²，
建筑面积61万m²
Planning & Architectural Design of Software
Industry Base (section III & IV) of Shenzhen
High-tech Industrial Park
2009 Bidding project Land area: 14.9ha
Building area: 610000m²

深圳中信银行大厦建筑设计
2009年，竞标方案，用地面积0.44hm²，
建筑面积6.58万m²
Architectrual Design of China CITIC BANK
(shenzhen) Mansion
2009 Bidding project Land area: 0.44ha
Building area: 65800m²

南方科技大学和深圳大学新校区拆迁安置项目
—产业园区建筑设计
2009年，竞标方案，用地面积15.25hm²，建筑面积63.10万m²
Removing and Resettlement of New Campus of South
University of Science and Technology and Shenzhen
University - Planning & Architectural Design of Industry Park
2009 Bidding project Land area: 15.25ha
Building area: 631000m²

* 深圳地铁竹子林车辆段改扩建工程上盖建筑设计
2009年，中标方案，用地面积1.90hm²，
建筑面积4.42万m²
Architectural Design for Buildings above Zhuzilin
Metro Station of Shenzhen
2009 Winning project Land area: 1.90ha
Building area: 44200m²

深圳中广核大厦建筑设计
2008年，国际竞标，用地面积1.01hm²，
建筑面积18.85万m²，高度159.98m
Architecture Design of
China Guangdong Nuclear Power
Industry's Office Tower in Shenzhen
2008 International bidding Land area :1.01ha
Building area: 188500 m² Height:159.98m

深圳虚拟大学园院校产业化综合大楼建筑设计
2008年，用地面积0.78hm²，
建筑面积3.2万m²
Architecture Design of Complex Office of
Shenzhen Virtual University Park
2008 Land area: 0.78ha
Building area: 32000m²

深圳人才园建筑设计
2008年，用地面积3.61hm²，
建筑面积8.32万m²
Architecture Design of
Shenzhen Human Resouces Park
2008 Land area:3.61ha
Building area: 83200m²

珠海歌剧院建筑设计
2008年，国际竞标，用地面积42hm²，
建筑面积4.3万m²
Architectural Design of Zhuhai Opera House
2008 International bidding
Land area: 42ha
Building area: 43000m²

深圳航天国际中心建筑设计
2008年，国际竞标，用地面积1.05hm²，
建筑面积15万m²，高度241.2m
Architectural Design of
International Aerospace Center of Shenzhen
2008 International bidding Land area: 1.05ha
Building area: 150000m² Height: 241.2m

成都怡湖玫瑰湾建筑设计
2008年，用地面积12.57hm²，
建筑面积62.5万m²
Architectual Design of
Yihu Rose Bay Community, Chengdu
2008 Land area: 12.57ha
Building area: 625000m²

深圳市光明新区公明文化艺术和体育中心建筑设计
2008年，用地面积9.9hm²，
建筑面积5.09万m²
Architecture Design of
Gongming Culture, Art & Sports Center of
Guangming New Town, Shenzhen
2008 Land are: 9.9 ha
Building area: 50900m²

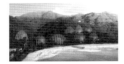

深圳南澳鹅公湾水产养殖基地3#地建筑设计
2008年，用地面积1.61hm²，
建筑面积0.64万m²
Architecture Design of
Aquiculture Base Plot 3 of Egong Bay in
Nan'ao, Shenzhen
2008 Land area: 1.61ha
Building area: 6400m²

深圳宝安中心区N28区九年一贯制学校建筑设计
2008年，用地面积2.7hm²，
建筑面积2.42万m²
Architecture Design of
Nine Year's Volunteer Educational school of
Bao'an Block N28, Shenzhen
2008 Land area: 2.7ha
Building area: 24200m²

深圳市龙岗区南湾中学建筑设计
2008年，用地面积2.75hm²，
建筑面积1.66万m²
Architecture Design of
Nine Year's Volunteer Educational school of
Longgang Nanwan in Shenzhen
2008, Land area: 2.75ha
Building area: 16600m²

宿州综合体概念性建筑设计
2008年，用地面积21.3hm²，
建筑面积67.92万m²
Architecture Design of Suzhou Complex
2008 Land area:21.3ha
Building area: 679200m²

* 顺德深业城一期建筑立面设计
2008年，建筑面积18万m²
Architecture Facade Design of
Shum Yip's Community (Phase 1), Shunde
2008 Building area: 180000m²

中国饮食文化城建筑设计
2008年，用地面积220hm²，
建筑面积38万m²
Architecture Design of
Chinese Gastrologic and Cultural Town of Shenzhen
2008 Land area: 220ha
Building area: 380000m²

深圳市南油购物公园建筑设计
2008年，国际竞标第一名，
用地面积13hm²，建筑面积45.4万m²
Architecture Design of
Nanyou Shopping Park, Shenzhen
2008, First prize of international bidding,
Land area: 13ha Building area: 454000m²

深圳光明新城万丈坡居住小区建筑设计
2008年，详蓝编制范围22hm²，
规划研究范围49.6hm²，建筑面积53.23万m²
Architecture Design of Wanzhangpo Residential
Community of Guangmin New Town, Shenzhen
2008 Planning area: 22ha
Research area: 49.6ha Building area: 532300m²

珠海五洲花城二期建筑设计
2008年用地面积16.5hm²，
建筑面积60万m²
Architecture Design of
Five Continental Residence Garden, Zhuhai
2008 Land area: 16.5ha
Building area: 600000m²

* 深圳康佳研发大厦建筑设计 方案一
2007年，中标方案，用地面积0.96hm²，
建筑面积7.84万m²，高度100m
Architecture Design of
Konka R&D Building, Shenzhen (Schema I)
2007 Winning project Land area: 0.96ha
Building area: 78400m² Height: 100m

深圳康佳研发大厦建筑设计 方案二
2007年，用地面积0.96hm²，
建筑面积7.93万m²，高度100m
Architecture Design of
Konka R&D Building, Shenzhen (Schema II)
2007 Land area: 0.96ha
Building area: 79300m² Height: 100m

* 深圳市长富金茂大厦建筑设计
2007年，中标方案，用地面积1.88hm²，
建筑面积22.19万m²，高度350m
Architecture Design of
Changfu Jinmao Building, Shenzhen
2007 Winning project Land area: 1.88ha,
Building area: 221900m² Height: 350m

* 深圳龙岗区天安数码新城建筑设计
2007年，中标方案，用地面积4.63hm²，
建筑面积24.68万m²
Architecture Design of
Longgang District Cyber Park, Shenzhen
2007 Winning project Land area: 4.63ha，
Building area: 246800m²

狮山客运站建筑设计
2007年，中标方案，用地面积9.5hm²，
建筑面积11.45万m²
Architecture Design of
Shishan Long Distance Bus Station
2007 Winning project Land area: 9.50ha,
Building area: 114500m²

深业集团惠州开发总部建筑方案设计
2007年，用地面积0.9hm²，
建筑面积7.93万m²
Architecture Design of Huizhou Development
Headquarters of Shum Yip Group
2007 Land area: 0.9ha
Building area: 79300m²

中国太湖文化论坛会议中心建筑方案设计
2007年，用地面积2.5hm²，
建筑面积2.94万m²
Architecture Design of
Convention Center of Taihu Lake Cultural Forum
2007 Land area: 2.5ha
Building area: 29400m²

* 深圳市贝丽中学建筑方案设计
2007年，用地面积1.95hm²，
建筑面积2.53万m²
Architecture Design of
Shenzhen Beili School
2007 Land area: 1.95ha
Building area: 25300m²

* 贵阳中天世纪新城中心商业建筑设计
2007年，用地面积5.05hm²，
建筑面积9.7万m²
Architecture Design of the Central Commerce of
Guiyang Zhongtian New Century City
2007 Land area: 5.05ha
Building area: 97000m²

* 成都华润置地翡翠城小学建筑方案设计
2007年，用地面积1.65hm²，
建筑面积1.71万m²
Architecture Design of
CRL Jade-City Primary School, Chengdu
2007 Land area: 1.65ha
Building area:17100m²

* 深圳市半岛城邦住宅区二期售楼中心
建筑概念方案设计
2007年，用地面积2000m²，
建筑面积1500m²
Architecture Design of Sales Center of
The Peninsula (Phase 2), Shenzhen
2007 Land area: 2000m²
Building area: 1500m²

* 深圳市半岛城邦二期配套小学
建筑概念方案设计
2007年，用地面积1.04hm²，
建筑面积1.02万m²
Architecture Design of
Primary School in The Peninsula (Phase 2), Shenzhen
2007 Land area: 1.04ha
Building area: 10200m²

* 成都华润置地二十四城小学建筑方案设计
2007年，用地面积1.94hm²，
建筑面积1.53万m²
Architecture Design of
Primary School of CRL 24th City, Chengdu
2007 Land area: 1.94ha
Building area: 15300m²

* 万科宁波金色水岸会所建筑设计
2007年，用地面积2000m²，
建筑面积4500m²，竣工时间2008年
Architecture Design of Club-House of
Vanke Golden Waterfront Community, Ningbo
2007 Land area: 2000m²
Building area: 4500m² Constructed in 2008

佛山南海区狮山镇文化体育公园建筑设计
2007年，竞标方案，用地面积13.9hm²，
建筑面积28万m²
Architecture Design of Shishan Cultural &
Sport Park, Nanhai District, Foshan
2007 Bidding scheme Land area: 13.9ha
Building area: 280000m²

* 深圳市城建集团观澜居住区规划建筑设计
2007年，中标方案，用地面积16.99hm²，
建筑面积34.6万m²
Architecture Design of
Guanlan Residential Community, Shenzhen
2007 Winning project Land area: 16.99ha
Building area: 346000m²

* 华润置地成都翡翠城四期公共建筑设计
2007年，用地面积7.56hm²，
建筑面积41.56万m²
Architecture Design of Public Buildings of
CRL Jade-City (Phase 4), Chengdu
2007 Land area: 7.56ha
Building area: 415600m²

* 深圳市南山商业文化中心区超高层建筑设计
2006年，国际竞标中标方案，用地面积2.57hm²，
建筑面积16.3万m²，高度288m
Architecture Design of the Tower in Nanshan
Commercial & Cultural Center, Shenzhen
2006 Winning project of international bidding
Land area: 2.57ha
Building area: 163000m² Height: 288m

深圳市东海中心建筑设计
2006年，用地面积3.3hm²，
建筑面积40.3万m²，建筑高度230m
Architecture Design of
Shenzhen Donghai Center
2006 Land area: 3.3ha
Building area: 403000m² Height: 230m

南宁市城建档案馆（新馆）建筑设计
2006年，用地面积1.21hm²，
建筑面积5.48万m²
Architecture Design of
Nanning Urban Construction Archives
2006 Land area: 1.21ha
Building area: 54800m²

深圳鹏基龙电工业城建筑设计
2006年，用地面积3.08hm²，
建筑面积8.64万m²
Architecture Design of
Pengji Longdian Office Block, Shenzhen
2006 Land area: 3.08ha
Building area: 86400m²

* 深圳市沙河世纪假日广场建筑设计
2006年，用地面积1.13hm²，
建筑面积10.24万m²，竣工日期：2009年
Architecture Design of
Shahe Century Holiday Plaza, Shenzhen
2006 Land area: 1.13ha,
Building area: 102400m² Constructed in 2009

四川文化城建筑设计
2006年，用地面积1.05hm²，
建筑面积11.4万m²
Architecture Design of Sichuan Cultural City
2006, Land area: 1.05ha,
Building area: 114000m²

无锡新世界国际纺织服装市场中心商务区概念性设计
2006年，用地面积8.78hm²，
建筑面积27.6万m²
Architecture Design of Central Commercial Area of
Wuxi New World International Textile and
Clothing Market
2006 Land area: 8.78ha
Building area: 276000m²

三洋厂房改造项目概念性设计方案
2006年，用地面积1.03hm²，
建筑面积1.65万m²
Architecture Design of
SANYO Workshop Rebuild Project
2006 Land area: 1.03ha
Building area: 16500m²

* 深圳华侨城中学（高中部）建筑设计
2005年，用地面积6.3hm²，
建筑面积3.32万m²，竣工日期：2008年
Architecture Design of
Shenzhen Bay High School
2005 land area: 6.3ha
Building area: 33200m² Constructed in 2008

杭州市钱江创业中心建筑设计
2005年，用地面积4.52hm²，
建筑面积11万m²
Architecture Design of
Qianjiang Creation Center, Hangzhou
2005, Land area: 4.52ha,
Building area: 110000m²

杭州钱江新城中央商务区办公建筑概念设计
2005年，用地面积2.71hm²，
建筑面积12.45万m²
Architecture Design of
Qianjiang New CBD Office, Hangzhou
2005, Land area: 2.71ha,
Building area: 124500m²

杭州天际大厦建筑设计
2005年，国际竞标，用地面积1.13hm²，
建筑面积5.25万m²
Architecture Design of
Tianji Mansion, Hangzhou
2005 International bidding Land area: 1.13ha
Building area: 52500m²

深圳市招商华侨城尖岗山商业中心建筑设计
2005年，国际竞标，用地面积6.86hm²，
建筑面积6.53万m²
Architecture Design of
OCT Jiangganshan Shopping Center, Shenzhen
2005 International bidding Land area: 6.86ha
Building area: 65300m²

* 上海浦南中学建筑设计
2005年，用地面积1.7hm²，
建筑面积1.06万m²
Architecture Design of
Punan High School, Shanghai
2005 Land area: 1.7ha
Building area: 10600m²

无锡博物馆、科技馆、革命陈列馆综合体建筑设计
2005年，国际竞标，
用地面积2.02hm²，建筑面积4.30万m²
Architecture Design of Wuxi Complex Museum
2005 International bidding
Land area: 2.02ha Building area: 43000m²

海与建筑-餐饮中心建筑设计意向
2005年，国际竞标，用地面积0.10hm²，
建筑面积570m²
Architecture Design of Sea & Building Restaurant
2005 International bidding Land area: 0.1ha,
Building area: 570m²

韶关风度广场建筑设计
2004年，用地面积3.2hm²，
建筑面积10.25万m²
Architecture Design of
Shaoguan Feng-Du Plaza
2004 Land area: 3.2ha
Building area: 102500m²

* 深圳市南山商业文化中心区11-01地块项目概念设计
2004年，用地面积2.57hm²，
建筑面积17.53万m²
Architecture Design of
No.11-01 Plot of Nanshan Business &
Culture Center, Shenzhen
2004 Land area: 2.57ha
Building area: 175300m²

* 成都置信天府花城建筑设计
2004年，用地面积3.03hm²，
建筑面积1.4万m²，竣工日期：2007年
Architecture Design of
Zhixin Tianfu Flower City Residence
2004 Land area: 3.03ha
Building area: 14000m² Constructed in 2007

* 成都置信未来广场A期建筑设计
2004年，用地面积6.39hm²，
建筑面积10.87万m²，竣工日期：2005年
Architecture Design of
Zhixin "Future Plaza" (Phase A), Chengdu
2004 Land area: 6.39ha
Building area: 108700m² Constructed in 2005

广东省南海盐步信息发展中心建筑设计
2004年，用地面积6hm²，
建筑面积3万m²
Architecture Design of Yanbu
Information Development Center, Nanhai
2004 Land area: 6ha
Building area: 30000m²

深圳市宝安城区26区商业公园建筑设计
2004年，规划设计面积26.7hm²，
总建筑面积65万m²
Architecture Design of
Commercial Park of Bao'an 26# Block, Shenzhen
2004 Planning area: 26.7ha
Building area: 650000m²

成都川投置信广场规划与建筑设计
2004年，用地面积0.84hm²，
建筑面积5.89万m²
Architecture Design of
Chuantou Zhixin Plaza, Chengdu
2004 Land area: 0.84ha
Building area: 58900m²

深圳市福田科技广场规划与建筑设计
2004年，用地面积3.87hm²，
建筑面积21.7万m²
Architecture Design of
Futian Science & Technology Plaza, Shenzhen
2004 Land area: 3.87ha
Building area: 217000m²

* 深圳欧贝特卡系统科技有限公司新工厂改造项目
2004年，总改造面积3975m²
Reconstruction Project of
New Workshop of Shenzhen Obetaca
System Science & Technology Co., Ltd
2004 Total reconstruction area: 3795m²

北京宣武区国信大吉片建筑设计
2004年，用地面积43.9hm²，
建筑面积165万m²
Architecture Design of Guoxin Dajipian of
Xuanwu District, Beijing
2004 Land area: 43.90ha
Building area: 1650000m²

湖南天健长沙芙蓉中路项目居住区建筑设计
2004年，用地面积17.12hm²，
建筑面积75.12万m²
Architecture Design of Tijian Residence on
Furongzhong Road, Changsha
2006 Land area: 17.12ha
Building area: 751200m²

* 深圳鹏基集团商务时空建筑设计
2003年，用地面积1.1hm²，
建筑面积2.7万m²，竣工日期：2007年
Architecture Design of
Pengji Business Space-time Mansion, Shenzhen
2003 Land area: 1.1ha,
Building area: 27000m² Constructed in 2007

海南省三亚阳光海岸城市建筑设计
2003年，国际竞标，
总设计面积146.67hm²
Architecture Design of
Sanya Sunny-Coast City, Hainan
2003 International bidding
Total design area: 146.67ha

* 深圳龙岗下沙金沙滩滨海休闲带建筑设计
2003年，用地面积17hm²，
竣工日期：2004年
Architecture Design of
Xiasha Golden Bench Seaside Resorts
2003 Land area: 17ha
Constructed in 2004

福建晋江市行政中心建筑设计
2003年，用地面积24hm²，
建筑面积8.5万m²
Architecture Design of
Jinjiang Administration Center, Fujian
2003 Land area: 24ha
Building area: 85000m²

郑州郑东新区颐和医院建筑设计
2003年，用地面积18.07hm²，
建筑面积20.62万m²
Architecture Design of Yihe Hospital, Zhengzhou
2003 Land area: 18.07ha
Building area: 206200m²

郑州郑东新区房地产展销大楼建筑设计
2003年，国际竞标第一名，
建筑面积6.5万m²
Architecture Design of Housing Display and Selling
Building of New East District, Zhengzhou
2003 First prize of international bidding
Building area: 65000m²

深圳星河万豪五星级酒店建筑设计
2003年，用地面积0.90hm²，
建筑面积11万m²
Architecture Design of
Xinghe Marriot International Hotel, Shenzhen
2003 Land area: 0.90ha
Building area: 110000m²

中央音乐学院珠海分校建筑设计
2003年，用地面积43hm²，
建筑面积15万m²
Architecture Design of
National Music University- Zhuhai Branch
2003 Land area: 43ha
Building area: 150000m²

深圳市龙岗区规划展览综合大楼建筑设计
2003年，国际竞标中标方案，
用地面积4.88hm²，建筑面积3.14万m²
Architecture Design of
Longgang Planning Exhibition Complex Building, Shenzhen
2003 Winning project of international bidding
Land area: 4.88ha Building area: 31400m²

* 深圳大剧院外观改造建筑设计
2003年，国际竞标中标方案，
总建筑面积8.52万m²，竣工日期：2006年
Architectural Facade Design of
Shenzhen Grand Theatre
2003 Winning project of international bidding
Building area: 85200m² Constructed in 2006

上海法德国际学校校园建筑设计
2003年，用地面积5.7hm²，
建筑面积2.50万m²
Architecture Design of
Shanghai French-German International School
2003 Land area: 5.7ha
Building area: 25000m²

杭州广厦-邮政大厦建筑设计
2003年，用地面积1.96hm²，
建筑面积9万m²
Architectue Design of Hangzhou Post Tower
2003 Land area: 1.96ha
Building area: 90000m²

* 北京金世纪酒店建筑立面设计
2003年，建筑面积3.50万m²，
竣工日期：2006年
Architectural facade Design of
Beijing Golden Century Hotel
2003 Building area: 35000m²
Constructed in 2006

杭州浙江宾馆四星级酒店建筑设计
2003年，用地面积9hm²，
建筑面积7万m²
Architecture Design of
Zhejiang Hotel (Four-star), Hangzhou
2003 Land area: 9 ha
Building area: 70000m²

浙江电子信息大厦建筑设计
2003年，国际竞标中标方案，
用地面积0.88hm²，建筑面积7万m²
Architecture Design of
Information Mansion, Zhejiang
2003 Winning project of international bidding
Land area: 0.88ha Building area: 70000m²

杭州下沙开发区城市建筑设计
2003年，用地面积85hm²，
建筑面积91万m²
Architecture Design of
Xiasha District, Hangzhou
2003 Land area: 85 ha
Building area: 910000m²

深圳葵涌海滨度假中心建筑设计
2002年，国际竞标，
建筑面积1.5万m²
Architecture Design of
Kuiyong Seaside Holiday Center, Shenzhen
2002 International bidding
Building area: 15000m²

* 广州法国文化协会（Alliance French）室内设计
2002年，设计面积215m²，
竣工日期：2002年
Interior Design of
Guangdong Alliance French
2002 Design area: 215m²
Constructed in 2002

* 深圳半岛城邦建筑设计
2002年，用地面积5hm²，
建筑面积18万m²，竣工日期：2007年
Architecture Design of The Peninsula, Shenzhen
2002 Land area: 5 ha Building area: 180000m²
Constructed in 2007

* 成都新城市广场建筑设计
2002年，建筑面积17万m²，
竣工日期：2005年
Architecture Design of Chengdu New City Plaza
2002 Building area: 170000m²
Constructed in 2005

* 杭州南北商务港建筑设计
2002年，建筑面积3万m²，
获奖项目，竣工日期：2005年
Architecture Design of
Nanbei Business Port, HangZhou
2002 Building area: 30000m²
Award project Constructed in 2005

成都国际商城建筑设计
2002年，国际竞标中标方案，
建筑面积18万m²，竣工日期：2006年
Architecture Design of
Chengdu International Shopping Mall
2002 Winning project of international bidding
Building area: 180000m² Constructed in 2006

* 广州新世界地产凯旋新世界建筑设计
2001年，用地面积10hm²，竣工日期：2004年
Architecture Design of Guangzhou Triumph
New World Residential Community
2001 Land area:10 ha Constructed in 2004

深圳市党委建筑设计
2001年，建筑面积8万m²
Architecture Design of Shenzhen
Municipal Party Committee and Party School
2001 Building area: 80000m²

深圳国际网球中心俱乐部小区建筑设计
2001年，用地面积9.8hm²，住宅建筑面积14万m²，
国际网球中心建筑面积3万m²
Architecture Design of Shenzhen International
Tennis Center Club Residence
2001 Land area: 9.8ha
Building area of house: 140000m²
Building area of international tennis center: 30000m²

深圳高新技术开发区中心区建筑设计
2001年，用地面积25.6hm²，
建筑面积27万m²
Architecture Design of Central Area of
Shenzhen Hi-tech Development Zone
2001 Land area: 25.6ha
Building area: 270000m²

深圳国际市长大厦建筑设计
2001年，建筑面积3.20万m²
Architectue Design of
Shenzhen International Mayors Building
2001 Building area: 32000m²

深圳华侨城中央会所建筑设计
2001年，建筑面积7000m²
Architecture Design of OCT
Central Club House, Shenzhen
2001 Building area: 7000m²

* 深圳华侨城中央科教所附属9年制学校建筑设计
2001年，国际竞标中标方案，用地面积3.30hm²，
建筑面积2.30万m²，竣工日期：2003
Architecture Design of Nanshan School Affiliated with
China National Institute for Educational Research
2001 Winning project of international bidding,
Land area: 3.3ha Building area: 23000m²
Constructed in 2003

* 深圳市梧桐山电视发射塔建筑设计
2001年，国际竞标中标方案，
建筑面积1万m²，高度300m
Architecture Design of
Wutong Mountain TV Transmission Tower, Shenzhen
2001 Winning project of international bidding
Building area: 10000m² Height: 300m

温州苍南新市中心区政务中心建筑设计
2001年，规划面积454hm²，
建筑面积12万m²
Architecture Design of Administration
Center of New Cangnan CBD, Wenzhou
2001 Planning area: 454ha
Building area: 120000m²

杭州市钱塘江两岸新中心区建筑设计
2001年，国际竞标，用地面积300hm²
Architecture Design of New CBD along the
Qiantangjiang River, Hangzhou
2001 International bidding Land area: 300 ha

济南市南部新城区政务中心建筑设计
2001年，规划面积396hm²，
建筑面积15万m²
Architecture Design of Administration
Center of New South District, Jinan
2001 Planning area: 396 ha,
Building area: 150000m²

深圳创智集团办公大楼建筑设计
2001年，建筑面积3.2万m²
Architecture Design of
Chuangzhi Group Office Building, Shenzhen
2001 Building area: 32000m²

南海市海八路以北建筑设计
2001年，用地面积174hm²，
建筑面积23.90万m²
Architecture Design of
the North of Haiba Road, Nanhai
2001 Land area: 174 ha
Building area: 239000m²

吉林市松花江沿岸总体规划构思及重点地区城市建筑设计
2001年，国际竞标，规划陆地面积2525hm²，
城市设计范围971hm²
Architecture Design of the Key Area of
Songhua River Shores, Jilin
2001 International bidding
Land area: 2525ha Urban design area: 971ha

杭州西湖文化广场建筑设计
2001年，建筑面积20万m²
Architecture Design of
West Lake Cultural Plaza, Hangzhou
2001 Building area: 200000m²

* 杭州嘉华国际中心办公楼建筑设计
2001年，建筑面积5.6万m²，
获全国人居典金奖、商务类金奖，竣工日期：2003年
Architecture Design of
Jiahua International Office Building , Hangzhou
2001 Building area: 56000m²
Awarded as the Golden Commercial Housing of
China   Constructed in 2003

法国Decathlon公司深圳基地项目建筑设计
2000年，可行性研究，用地面积3hm²
Architecture Design of
Decathlon (France) Shenzhen Office
2000 Feasibility study
Land area: 3 ha

深圳福田新中心区香港凤凰卫视亚洲总部建筑设计
2000年，可行性研究，建筑面积8万m²
Architecture Design of
HongKong Phoenix TV's Aisa Headquarter
in Futian CBD, Shenzhen
2000 Feasibility study
Building area: 80000m²

* 安徽合肥国际会议展览中心建筑设计
2000年，
建筑面积5.6万m²，竣工日期：2002年
Architecture Design of Hefei International
Conference & Exhibition Center, Anhui
2000 Winning project of international bidding
Building area: 56000m² Constructed in 2002

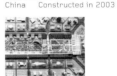
深圳宝安中心城建筑设计
2000年，国际竞标，用地面积100hm²，
建筑面积8.3万m²，绿化面积30万m²
Architecture Design of Central District of Bao'an,
Shenzhen
2000 International bidding Land area: 100ha,
Building area: 83000m² Green area: 300000m²

* 深圳盐田沙头角体育中心建筑设计
2000年，建筑面积8000m²，
竣工日期：2004年
Architecture Design of Shatoujiao
Sports Center of Yantian District, Shenzhen
2000 Building area: 8000m²
Constructed in 2004

深圳星昱大厦办公楼建筑设计
2000年，建筑面积4.5万m²
Architecture Design of
Xingyan Office Building, Shenzhen
2000 Building area: 45000m²

湖州市仁皇山新区城市建筑设计
2000年，国际竞标，用地面积400hm²
Architecture Design of
Renhuangshan New District, Huzhou
2000 International bidding
Land area: 400ha

* 深圳规划国土局盐田分局办公楼建筑设计
1999年，国际竞标中标方案，用地面积0.6hm²
建筑面积9469m²，竣工日期：2001年
Architecture Design of Yantian Branch Office
Building of Shenzhen Urban Planning & Land Bureau
1999 Winning project of international bidding
Land area: 0.6 ha Building area: 9469m²
Constructed in 2001

* 深圳规划国土局盐田分局办公楼室内设计
1999年，设计面积7000m²，
竣工日期：2001年
Interior Design of Yantian Branch Office Building
of Shenzhen Urban Planning & Land Bureau
1999 Design area: 7000m²
Constructed in 2001

深圳地王大厦第68层凌霄阁室内设计
1999年，设计面积2000m²，
竣工日期：2003年
Interior Design of Lingxiaoge on the 68th floor of
Diwang Commercial Center
1999 Design area: 2000m²
2000 Constructed in 2003

杭州市大剧院建筑设计
1999年，国际竞标，
建筑面积4.5万m²
Architecture Design of Hangzhou Grand Theatre
1999 International bidding
Building area: 45000m²

杭州市吴山商城建筑设计
1999年，国际竞标，用地面积6hm²，
建筑面积9万m²
Architecture Facade Design of
Wushan Shopping Mall, Hangzhou
1999 International bidding
Land area: 6 ha Building area: 90000m²

深圳罗湖华佳广场建筑立面设计
1998年，竣工日期：1998年
Architecture Facade Design of
Shenzhen Luohu Huajia Plaza
1998 Constructed in 1998

* 珠海珠澳海关地区规划与口岸联检大楼建筑设计
1998年，规划用地面积60hm²，
建筑面积7万m²，竣工日期：1998年
Planning Design of Zhuhai Zhu'ao Custom Area and
Architecture Design of Port Checking Building
1998 Land area of planning: 60ha,
Building area: 70000m² Constructed in 1998

* 深圳华侨城OCT生态广场建筑设计
1998年，城市公园景观设计用地面积5hm²，
建筑面积3万m²，竣工日期：1999年
Architecture Design of OCT Ecology Plaza, Shenzhen
1998 Land area of landscape design: 5 ha
Building area: 30000m² Constructed in 1999

深圳市文化广场（现更名为中信广场）建筑设计
1998年，国际竞标，建筑面积15万m²
Architecture Design of Shenzhen Cultural Plaza
(Zhongxin City Plaza now)
1998 International bidding
Building area: 150000m²

江苏省南京市师范大学校园区建筑设计
1997年，国际竞标二等奖，
建筑面积25万m²
Architecture Design of Nanjing
Normal University, Jiangsu
1997 Second prize of international bidding,
Building area: 250000m²

北京市乐邦集团和中国电影合作制片公司
综合大厦建筑设计
1997年，建筑面积5万m²
Architecture Design of Integrated Building of
Beijing Lebang Group and
China Film Co-product Corporation
1997 Building area: 50000m²

深圳电视中心工程建筑设计
1997年，国际竞标中标方案，
建筑面积4.5万m²
Architecture Design of Shenzhen TV Tower
1997 Winning project of international bidding
Building area: 45000m²

* 深圳市华侨城中西部城市综合区城市建筑设计
1996年，城市和环境设计用地面积85hm²，
建筑面积185万m²
Architecture Design of
Shenzhen OCT Central-West Area
1996 Land area of urban and
environmental design: 85ha
Building area: 1850000m²

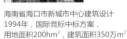

深圳市福田新市中心和市民中心建筑设计
1996年，国际竞标，用地面积200m²
市政厅建筑面积10万m²
Architecture Design of Futian New City Center and
Citizen Center of Shenzhen
1996 International bidding
Land area: 200ha Building area of city hall: 100000m²

上海东方医院建筑设计
1995年，国际竞标二等奖，
建筑面积4.5万m²
Architecture Design of Shanghai Oriental Hospital
1995 Second prize of international bidding
Building area: 45000m²

浙江省杭州国际金融中心建筑设计
1994年，国际竞标二等奖，
建筑面积5万m²
Architecture Design of Hangzhou
International Finance Center, Zhejiang
1994 Second prize of international bidding
Building area: 50000m²

海南省海口市新城市中心建筑设计
1994年，国际竞标中标方案，
用地面积200hm²，建筑面积350万m²
Architecture Design of New City Center of
Haikou, Hainan
1994 Winning project of international bidding
Land area: 200ha Building area: 3500000m²

珠海横琴行政中心建筑设计
1994年，用地面积100hm²
Architecture Design of
Zhuhai Hengqin Administration Center
1994 Land area: 100ha

北京市长安街皇城内珠宝商业中心设计
1994年，国际竞标一等奖，
建筑面积4.5万m²
Architecture Design of Jewelry Industry Center in
Imperial City of Chang'an Street, Beijing
1994 First prize of international bidding
Building area: 45000m²

上海东方音乐厅建筑设计
1994年，国际竞标二等奖，
建筑面积4万m²
Architecture Design of Shanghai Oriental Odeum
1994 Second prize of international bidding
Building area: 40000m²

* 珠海大学校园建筑设计
1993年，国际竞标一等奖，用地面积176hm²，
建筑面积36万m²，竣工日期：2000年
Architecture Design of Zhuhai University
1993 First prize of international bidding
Land area: 176ha Building area: 360000m²
Constructed in 2000

住宅
Residence

* 已竣工或建造中项目
* Projects constructed or under construction

* 成都天府华侨城纯水岸C组团高层区及公共配套项目建筑设计
2009年，委托设计，用地面积2.51hm²，建筑面积11.23万m²
Architectural Design of Residence Towers &
Public Supporting Facilities of Block C of Chengdu
OCT-The Riviera
2009 Mandated project Land area: 2.51ha
Building area: 112300m²

* 深圳城建集团观澜项目建筑设计
2009年，中标方案，用地面积19.46hm²，
建筑面积55.53万m²
Architectural Design of SZEG Guanlan
Residential Community in Shenzhen
2009 Mandated project Land area: 19.46ha
Building area: 555300m²

* 中山金源花园住宅建筑设计
2009年，中标方案，用地面积21.22hm²，
建筑面积79.26万m²
Architectural Design of Jin Yuan Garden
Residence in Zhongshan
2009 Winning project Land area: 21.22ha
Building area: 792600m²

南方科技大学和深圳大学新校区拆迁安置项目-
商住综合区住宅建筑设计
2009年，国际竞标招标方案，用地面积17.77hm², 建筑面积60万m²
Removing and Resettlement of New Campus of South University
of Science and Technology and Shenzhen University -
Architectural Design for Commerce and Residence
2009 Bidding project Land area: 17.77ha
Building area: 600000m²

★ 深圳半岛城邦3期超高层住宅建筑设计
2009年，委托设计，用地面积5.68hm²，
建筑面积19.07万m²
Architectural Design of Mega High Rise
Residential Tower of Shenzhen Peninsula (phase III)
2009 Mandated project Land area: 5.68ha
Building area: 190700m²

长春市净月区梧桐街住宅建筑设计
2008年，用地面积34.8hm²，
建筑面积45万m²
Architecture Design of Residential Community of
Phoenix Tree Street of Jingyue District, Changchun
2008 Land area: 34.8ha
Building area: 450000m²

深圳市光明新区保障性住房工程住宅建筑设计
2008年，用地面积4.12hm²，
建筑面积13万m²
Architecture Design of Social Community of
Guangming New Town, Shenzhen
2008 Land area: 4.12ha
Building area: 130000m²

成都怡湖玫瑰湾住宅建筑设计
2008年，用地面积12.57hm²，
建筑面积62.5万m²
Architecture Design of Community of
Yihu Rose Bay, Chengdu
2008 Land area: 12.57ha
Building area: 625000m²

长春市高新区C-6地块住宅建筑设计
2008年，用地面积8.57hm²，
建筑面积15.97万m²
Architecture Design of Residential Community of
High-Tech C-6 Block, Changchun
2008 Land area: 8.57ha
Building area: 159700m²

宿州综合体概念性建筑设计
2008年，用地面积21.3hm²，
建筑面积67.92万m²
Architecture Design of Suzhou Complex
2008 Land area: 21.3ha
Building area: 679200m²

★ 顺德深业城一期住宅建筑立面设计
2008年，建筑面积18万m²
Architectural Facade Design of
Shum Yip's Community (Phase 1), Shun
2008 Building area: 180000m²

中国饮食文化城建筑设计
2008年，用地面积220hm²，
建筑面积38万m²
Architecture Design of
Chinese Gastrologic and Culture Town, Shenzhen
2008 Land area: 220ha
Building area: 380000m²

珠海五洲花城二期住宅建筑设计
2008年，用地面积16.5hm²，
建筑面积60万m²
Architecture Design of
Five Continental Residence Garden, Zhuhai
2008 Land area: 16.5ha
Building area: 600000m²

深物业彩天怡色家园建筑设计
2008年，用地面积0.51hm²，
建筑面积3.6万m²
Architecture Design of
'Colourful Sky' Residence, Shenzhen
2008 Land area: 0.51ha
Building area: 36000m²

成都麓山国际超高层住宅建筑设计
2008年，用地面积3.61hm²，
建筑面积12万m²
Architecture Design of Super High-Rise Reside
Tower of Lushan International Community, Chen
2008 Land area: 3.61ha
Building area: 120000m²

成都天府华侨城二期住宅建筑设计
2008年，用地面积10hm²，
建筑面积15万m²
Architecture Design of
Tianfu OCT Residence (Phase 2), Chengdu
2008 Land area: 10ha
Building area: 150000m²

中山市坦州镇居住小区住宅建筑设计
2008年，用地面积11.4hm²，
建筑面积28万m²
Architecture Design of
Tanzhou Residential Community, Zhongshan
2008 Land area: 11.4ha
Building area: 280000m²

★ 贵阳中天世纪新城三期四组团建筑设计
2007年，用地面积5.9hm²，
建筑面积23.15万m²
Architecture Design of the 4# Block of
Zhongtian New Century City (Phase 3), Guiyang
2007 Land area: 5.9tha ,
Building area: 231500m²

★ 西安高科城市风景8#府邸建筑设计
2007年，用地面积2.6hm²，
建筑面积9.10万m²
Architecture Design of
the 8# Mansion of Gaoke City View, Xi'an
2007 Land area: 2.6ha
Building area: 91000m²

深业·坪山居住小区规划与建筑方案设计
2007年，用地面积2.83hm²，
建筑面积11.71万m²
Architecture Design of Pingshan
Residential Community
2007 Land area: 2.83ha
Building area: 117100m²

珠海城市壹站居住区规划与建筑方案设计
2007年，用地面积2.85hm²，
建筑面积8.31万m²
Architecture Design of The First City Station
Residential Community, Zhuhai
2007 Land area: 2.85ha
Building area: 83100m²

★ 华润置地成都翡翠城四期住宅立面设计
2007年，用地面积7.56hm²，
建筑面积41.56万m²
Architectural Facade Design of
CRL Jade-City Residence (Phase 4)
2007 Land area: 7.56ha
Building area: 415600m²

佛山南海区狮山镇文化体育公园住宅设计
2007年，国际竞标，用地面积13.9hm²，
建筑面积28万m²
Architecture Design of the Residence of Shishan
Culture& Sport Park, Nanhai district, Foshan
2007 International bidding Land area: 13.9ha
Building area: 280000m²

* 深圳市城建集团观澜居住区规划建筑设计
2007年，国际竞标中标方案，
用地面积16.99hm², 建筑面积34.6万m²
Architecture Design of Guanlan
Residential Community, Shenzhen
2007 Winning project of international bidding
Land area: 16.99ha Building area: 346000m²

* 鹏基惠州半山名苑居住建筑设计
2007年，用地面积2.6hm²,
建筑面积9.1万m², 竣工日期：2009年
Architecture Design of
Pengji Hillside Residential Community, Huizhou
2007 Land area: 2.60ha
Building area: 91000m² Constructed in 2009

江苏省姜堰市锦绣姜城规划与建筑方案设计
2007年，用地面积29.62hm²,
建筑面积62万m²
Architecture Design of Pengji
Splendid City, Jiangyan, Jiangsu Province
2007 Land area: 29.62ha
Building area: 620000m²

武汉万科高尔夫城市花园规划建筑设计
2007年，用地面积13.56hm²,
建筑面积27.9万m²
Architecture Design of
Vanke Golf City Garden Residence, Wuhan
2007 Land area: 13.56ha
Building area: 279000m²

深圳天鹅堡三期前期研究
2007年，用地面积11.17hm²,
建筑面积25.9万m²
Preliminary Studies for
The Swan Castle of OCT, Shenzhen
2007 Land area: 11.17ha
Building area: 259000m²

深圳中信惠州东江新城一期项目规划与建筑设计
2007年，用地面积25hm²,
建筑面积60万m²
Architecture Design of
CITIC Dongjiang New City (Phase 1), Huizhou
2007 Land area: 25ha
Building area: 600000m²

* 贵阳中天世纪新城三组团规划与建筑设计
2006年，用地面积17.54hm²,
建筑面积23.68万m²
Architecture Design of 3# Block of
Zhongtian New Century City, Guiyang
2006 Land area: 17.54ha
Building area: 236800m²

* 深圳蛇口君汇新天住宅小区规划与建筑设计
2006年，用地面积4.5hm²,
建筑面积11.65万m²
Architecture Design of
Junhuixintian Residential
Community, Shekou, Shenzhen
2006 Land area: 4.50ha
Building area: 116500m²

深圳市怡东花园规划建筑设计
2006年，用地面积6.1hm²,
建筑面积23.1万m²
Architecture Design of
Yidong Graden Residence, Shenzhen
2006 Land area: 6.10ha
Building area: 231000m²

* 合肥澜溪镇A、B区建筑及规划设计
2005年，用地面积7.23hm²,
建筑面积10.63万m²
竣工日期：2007年，获奖项目
Planning Design of Section A&B of Nancy Town, Hefei
2005 Land area: 7.23ha Building area: 106300 m²
Constructed in 2007 Awarded project

深圳市宝安26区旧城改造项目住宅设计
2005年，国际竞标，用地面积23.13hm²,
建筑面积80.7万m²
Architecture Desig for Renovation of
26# Block in Bao'an, Shenzhen
2005 International bidding Land area: 23.13ha
Building area: 807000m²

招商华侨城尖岗山商业中心住宅设计
2005年，国际竞标，用地面积6.86hm²,
建筑面积6.53hm²
Architecture Design of
OCT Jiangganshan Shopping Center, Shenzhen
2005 International bidding Land area: 6.86ha
Building area: 65300m²

长沙天际岭项目住宅区设计
2005年，国际竞标，用地面积38.81hm²,
建筑面积39.77万m²
Architecture Design of
Tianjiling Residential Community, Changsha
2005 International bidding
Land area: 38.81ha Building area: 397700m²

苏州工业园区"左岸山庭"住宅区设计
2005年，国际竞标，用地面积19.54hm²,
建筑面积39.07万m²
Architecture Design of "Hill-Yard on Left-Bank"
Residence, Suzhou
2005 International bidding Land area: 19.54ha
Building area: 390700m²

* 成都阳明山庄城市别墅住宅区设计
2005年，用地面积33.34hm²,
建筑面积26.56万m²
Architecture Design of Yangming
Shanzhuang City Villa, Chengdu
2005 Land area: 33.34ha
Building area: 265600m²

* 成都天府长城2期住宅建筑设计
2005年，中标方案，用地面积6.22hm²,
建筑面积21万m², 竣工日期：2007年
Architecture Design of
Tianfu Great Wall Residence (Phase 2), Chengdu
2005 Winning project Land area: 6.22ha
Building area: 210000m² Constructed in 2007

深圳市金光华"绿谷蓝溪"住宅小区设计
2005年，国际竞标，用地面积19.92hm²,
建筑面积35.84万m²
Architecture Design of Jinguanghua
"Lvgu Lanxi" Community, Shenzhen
2005 International bidding Land area: 19.92ha
Building area: 358400m²

北京密云休闲度假小区住宅设计
2005年，用地面积8hm²,
建筑面积14.28万m²
Architecture Design of
Beijing Miyun Vacation Residence
2005 Land area: 8ha
Building area: 142800m²

* 成都中海国际社区住宅设计
2004年，用地面积132.42hm²,
建筑面积39.6万m²
Design of Chengdu Zhonghai International
Community Residential
2004 Land area: 132.42ha
Building area: 3960000m²

* 华润置地成都翡翠城汇锦云天住宅建筑设计
2004年，国际竞标第二名，用地面积80hm²,
建筑面积86.65万m², 竣工日期：2007年
Architecture Design of
CRL Jade City Residence (Phase 2), Chengdu
2004, Second prize of international bidding
Land area: 80ha
Building area: 866500m² Constructed in 2007

北京宣武区国信大吉片住宅设计
2004年，用地面积43.9hm²，
建筑面积165万m²
Planning and Design of
Xuanwu District Guoxin Dajipian Residence, Beijing
2004 Land area: 43.90ha
Building area: 1650000m²

湖南天健长沙芙蓉中路项目居住区设计
2004年，用地面积17.12hm²，
建筑面积75.12万m²
Architecture Design of
"Tianjian Changsha Furongzhong Road", Hunan
2004 Land area: 17.12ha，
Building area: 751200m²

福州市登云山庄居住区设计
2004年，用地面积267.4hm²，
建筑面积68.16万m²
Architecture Design of
Dengyun Mountain Village, Fuzhou
2004 Land area: 267.40ha
Building area: 681600m²

深圳市宝安城区26区商业公园居住区设计
2004年，规划面积26.7hm²，
建筑面积65万m²
Architecture Design of
Bao'an 26#Block Commercial Park, Shenzhen
2004 Planning area: 26.7ha
Building area: 650000m²

西安中海华庭居住小区设计
2004年，用地面积5.41hm²，
建筑面积17.3万m²
Architecture Design of
Zhonghai Luxuriant Garden Residence, Xi'an
2004 Land area: 5.41ha，
Building area: 173000m²

深圳市鸿荣源龙岗中心城居住区设计
2004年，国际竞标，用地面积40hm²，
建筑面积60万m²
Architecture Design of Hongrongyuan
Longgang Central City, Shenzhen
2004 International bidding Land area: 40ha，
Building area: 600000m²

深圳市西乡富通住宅区设计
2004年，总用地面积25hm²，
建筑面积50万m²
Architecture Design of Futong Residential
Community of Bao'an District, Shenzhen
2004 Land area: 25ha
Building area: 500000m²

上海金地普罗旺斯风情街住宅设计
2004年，用地面积2.3hm²，
建筑面积1.90万m²
Architecture Design of the Residence of
Jindi Provence Street, Shanghai
2004 Land area: 2.30ha
Building area: 19000m²

* 深圳市恒立海岸花园居住区设计
2004年，用地面积2hm²，
建筑面积6万m²，竣工日期：2005年
Architecture Design of
Hengli Seashore Garden Residence, Shenzhen
2004 Land area: 2ha
Building area: 60000m² Constructed in 2005

* 深圳天琴湾项目三期别墅区单体设计
2004年，用地面积0.5hm²，
建筑面积1720m²
Architecture Design of
Lyra Bay Villa (Phase 3), Shenzhen
2004 Land area: 0.5ha，
Building area: 1720m²

深圳市沙河世纪山谷住宅小区设计
2003年，用地面积18hm²，
建筑面积67万m²
Architecture Design of
Shahe Century Valley Residential Community, Shenzhen
2003 Land area: 18ha
Building area: 670000m²

苏州新加坡工业园区东湖大郡二期住宅小区设计
2003年，用地面积10hm²，
建筑面积18万m²
Architecture Design of East Lake Residence
(Phase 2) of Singapore Industry Park, Suzhou
2003 Land area: 10ha
Building area: 180000m²

东莞世纪新城首期住宅小区设计
2003年，用地面积13hm²，
建筑面积18万m²
Architecture Design of
Center New City (Phase 1), Dongguan
2003 Land area: 13ha
Building area: 180000m²

* 苏州工业园区东城郡住宅建筑设计
2003年，用地面积6.5hm²，
建筑面积14.70万m²，竣工日期：2005年
Architecture Design of Residential Community of
East County of Suzhou Industry Park
2003 Land area: 6.50ha
Building area: 147000m² Constructed in 2005

上海西郊古北国际别墅设计
2003年，建筑面积602.32hm²
Architecture Design of
Gubei International Villa, Shanghai
2003 Building area: 602.32m²

* 成都新城市广场公寓设计
2002年，建筑面积17万m²
竣工日期：2005年
Architecture Design of
New City Plaza Apartment, Chengdu
2002 Building area: 170000m²,
Constructed in 2005

* 杭州盛德嘉苑三期住宅小区设计
2002年，建筑面积6.1万m²，
竣工日期：2005年
Architecture Design of
Shengdejiayuan (Phase 3), Hangzhou
2002 Building area: 61000m²,
Constructed in 2005

* 深圳半岛城邦1期住宅建筑设计
2002年，用地面积5hm²，建筑面积18万m²，
竣工日期：2007年 获奖项目
Architecture Design of
The Peninsula Residence (Phase 1), Shenzhen
2002 Land area: 5ha Building area: 180000m²
Constructed in 2007 Award project

上海嘉定高尔夫居住区设计
2002年，用地面积667hm²
Architecture Design of
Jiading Golf Residential Community, Shanghai
2002 Land area: 667ha

南京将军山别墅及多层住宅小区设计
2002年，用地面积16hm²，
建筑面积23.50万m²
Architecture Design of
Mt.General Villa and Multi-
floored Residence, Nanjing
2002 Land area: 16ha Building area: 235000m²

* 福州融侨"江南水都"首期住宅建筑设计
2002年，中标方案，用地面积12.34hm²，
建筑面积18.4万m²，竣工日期：2006年
Architecture Design of Rongqiao
"Jiang nan Water Town" Residence (Phase 1), Fuzhou
2002 Winning project Land area: 12.34ha
Building area: 184000m² Constructed in 2006

深圳国际网球中心俱乐部小区建筑设计
2001年，用地面积9.8hm²，住宅建筑面积14万m²，
国际网球中心建筑面积3万m²
Architecture Design of International
Tennis Center Club Residence, Shenzhen
2001 Land area: 9.8ha
Building area of house: 140000m², Building
area of international tennis center: 30000m²

* 贵阳中天世纪新城一期联排住宅小区建筑设计
2001年，用地面积8.4hm²，
建筑面积5.2万m²，竣工日期：2003年
Architecture Design of the Townhouse of
Zhongtian Century New City (Phase 1), Guiyang
2001 Land area: 8.40ha
Building area: 52000m² Constructed in 2003

成都长城地产五洲花园住宅区设计
2001年，用地面积63hm²，
建筑面积80万m²
Architecture Design of Changcheng Five
Continents Residence Garden, Chengdu
2001 Land area: 63ha
Building area: 800000m²

深圳万科下沙滨海住宅设计
2001年，建筑面积24万m²
Architecture Design of
Vanke Xisha Seashore Residence, Shenzhen
2001 Building area: 240000m²

深圳星河地产"叠翠九重天"住宅设计
2001年，用地面积59.71hm²，
建筑面积40万m²
Architecture Design of
Xinghe Real Estate "Rich Green Paradise"
Residential Community, Shenzhen
2001 Land area: 59.71ha
Building area: 400000m²

北京万科青青家园住宅小区设计
2000年，用地面积23hm²，
建筑面积28万m²
Architecture Design of
Vanke Qingqing House, Beijing
2000 Land area: 23ha
Building area: 280000m²

深圳华侨城中央花园住宅设计
2000年，建筑面积2.8万m²
Architecture Design of OCT Central Garden
Residence, Shenzhen
2000 Building area: 28000m²

杭州市吴山商城住宅设计
1999年，国际竞标，用地面积6hm²，
建筑面积9万m²
Architecture Design of the Residence of
Wushan Business City
1999 International bidding
Land area: 6ha Building area: 90000m²

山东省青岛世纪广场住宅设计
1999年，用地面积3hm²，
建筑面积6万m²
Architecture Design of
Century Plaza Residence, Qingdao
1999 Land area: 3ha
Building area: 60000m²

* 深圳盐田碧海名峰（现名天琴海）别墅区住宅设计
1998年，用地面积9hm²，
建筑面积3.8万m²
Architecture Design of Bihaimingfeng
(Lyra Bay) Villa, Yiantian District, Shenzhen
1998 Land area: 9ha
Building area: 38000m²

* 深圳世界花园第八期—海华居住宅设计
1998年，建筑面积15万m²，
竣工日期：2001年
Architecture Design of World Garden (Phase 8)
—Haihua Residence, Shenzhen
1998 Building area: 150000m²
Constructed in 2001

深圳市世界广场住宅设计
1998年，国际竞标中标方案，
建筑面积10.5万m²
Architecture Design of World Plaza Residence,
Shenzhen
1998 Winning project of international bidding
Building area: 105000m²

深圳观澜高尔夫球会会员住宅设计
1998年，建筑面积7.7万m²
Architecture Design of Guanlan Golf Club
VIP's Residence, Shenzhen
1998 Building area: 77000m²

深圳雅庭苑住宅小区建筑设计
1998年，国际竞标中标方案，
建筑面积10.4万m²
Architecture Design of Yating Residence, Shenzhen
1998 Winning project of international bidding
Building area: 104000m²

江苏省无锡法国达能矿泉水厂和生活基地设计
1997年，建筑面积3万m²
Architecture Design of French Daneng Mineral
Water Factory and Living Area, Wuxi, Jiangsu
1997 Building area: 30000m²

广西壮族自治区来宾法国电力公司厂区住宅设计
1997年，建筑面积3万m²
Architecture Design of Residential Area of Laibin
French Power Company Workshop, Guangxi
1997 Building area: 30000m²

江苏东海市圣戈班玻璃工厂和生活基地法国专家村设计
1997年，用地面积18hm²，建筑面积2.5万m²
Architecture Design of St. Gorban Glass
Manufacturer and Living Area French Expert
Village, Donghai, Jangsu
1997 Land area: 18ha
Building area: 25000m²

北京国门广场住宅区设计
1995年，建筑面积28万m²
Architecture Design of Beijing National Gate Plaza
Residential Community
1995 Building area: 280000m²

上海市浦东六里现代化生活居住区园区设计
1995年，建筑面积30万m²
Architecture Design of Pudong Six Li
Modern Life Residence Community, Shanghai
1995 Building area: 300000m²

景观
Landscape

* 已竣工或建造中项目
* Projects constructed or under construction

* 贵阳国际会议展览中心景观设计
2009年，委托设计，用地面积51.8hm², 景观面积31.81万m²
Landscape Design of Guiyang International Conference & Exposition Center
2009 Mandated project Land area: 51.8ha landscape area 318100m²

深圳南头古城——2街1园景观设计
2009年，竞标方案，用地面积10.8hm²
Landscape Design of Shenzhen Nantou Ancien City - 2 streets & 1 park
2009 Bidding project Land area: 10.8ha

泉州洛江现代世界农业旅游观光生态小镇景观概念深化设计
2009年，委托设计，用地面积510.8hm²
Conceptional Design of Ecological Town for Luojiang Modern Agricultural Tourism, Quanzhou
2009 Mandated project Land area: 510.8ha

* 深圳长富金茂大厦景观设计
2009年，委托设计，用地面积1.88hm²
Landscape Design of Changfu Jinmao Tower, Shenzhen
2009 Mandated project Land area: 1.88ha

* 中山金源花园景观设计
2009年，中标方案，用地面积21.22hm²
Landscape Design of Jin Yuan Garden Residence in Zhongshan
2009 Winning project Land area: 21.22ha

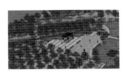

* 深圳城建集团碧中园二期景观设计
2009年，中标方案，用地面积2.5hm²，景观面积2.2万m²
Landscape Design of SZEG Bi Zhong Yuan Residence (phase II) in Shenzhen
2009 Winning project Land area: 2.5ha landscape area 22000m²

* 深圳城建集团观澜项目景观设计
2009年，中标方案，用地面积20m²，景观面积13万m²
Architectural Design of SZEG Guanlan Residential Community in Shenzhen
2009 Winning project Land area: 20ha landsacpe area:130000m²

* 深圳半岛城邦三期景观设计
2009年，委托设计，用地面积5.68hm²
Landscape Design of Shenzhen Peninsula (phase III)
2009 Mandated project Land area: 5.68ha

* 昆明滇池法国新站主题公园景观设计
2009年，竞标方案，用地面积100hm²，景观公园面积40万m²
Landscape Design of French New Station of Dian Lake of Kunming
2009 Bidding project Land area: 100 ha landscape area: 400000m²

深圳市宝安中心区银晖路商业街及核心商业区步行商业街景观设计
2008年，中标方案，用地面积4.95hm²
Landscape Design of Yinhui Shopping Street ar Pedestrian Shopping Street of Core Shopping Area in Bao'an Central District, Shenzhen
2008 Winning project Land area: 4.95ha

深圳人才园景观设计
2008年，用地面积3.61hm²
Landscape Design of Human Resouces Park, Shenzhen
2008 Land area: 3.61ha

* 深圳北站交通枢纽工程景观设计
2008年，中标方案，用地面积10hm²
Landscape Design of Transportation Hub of North Railway Station, Shenzhen
2008 Winning project
Land area: 10ha

珠海歌剧院景观设计
2008年，用地面积42hm²
Landscape Design of Zhuhai Opera House
2008 Land area: 42ha

成都怡湖玫瑰湾景观设计
2008年，用地面积12.57hm²
Landscape Design of Yihu Rose Bay Residentia Community, Chengdu
2008 Land area: 12.57ha

深圳市光明新区公明文化艺术和体育中心景观设计
2008年，用地面积9.9hm²
Landscape Design of Gongming Culture, Art & Sports Center of New Guangming Town, Shenzhen
2008 Land area: 9.9ha

泉州洛江农业旅游生态观光小镇景观概念设计
2008年，用地面积265hm²
Conceptual Landscape Design of Luojiang Ecological & Agricultural Tourist Town, Quanzhou
2008 Land area: 265ha

* 贵阳中天世纪新城中心商业组团景观设计
2008年，用地面积8.6hm²
Landscape Design of Central Commercial Block of Zhongtian Century New Town, Guiyang
2008 Land area: 8.6ha

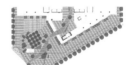

* 深圳康佳集团研发大厦景观设计
2008年，中标方案，用地面积0.96hm²
Landscape Design of Konka R&D Building, Shenzhen
2008 Winning project
Land area: 0.96ha

晋江新中心区市民广场景观概念设计
2008年，用地面积11hm²
Conceptual Landscape Design of
Citizen Plaza of New City Center of Jinjiang
2008 Land area: 11ha

＊ 成都中航城市广场景观设计
2008年，用地面积1.98hm²
Landscape Design of
CATIC City Plaza, Chengdu
2008 Land area: 1.98ha

成都中航国际广场景观设计
2008年，用地面积1.5hm²
Landscape Design of
CATIC International Plaza, Chengdu
2008 Land area: 1.5ha

中国饮食文化城景观设计
2008年，用地面积220hm²
Landscape Design of
Chinese Gastrologic and Culture Town, Shenzhen
2008 Land area: 220ha

深圳市南油购物公园景观设计
2008年，国际竞标第一名，
用地面积13hm²
Landscape Design of
Nanyou Shopping Park, Shenzhen
2008 First prize of International bidding
Land area: 13ha

珠海五洲花城二期概念性景观设计
2008年，用地面积16.5hm²
Landscape Design of
Five Continental Residence Garden, Zhuhai
2008 Land area: 16.5ha

＊ 深圳宝安区石岩大树林体育公园景观设计
2008年，中标方案，
用地面积5.1hm²
Landscape Design of
Shiyan Sporting Park of Bao'an District, Shenzhen
2008 Winning project
Land area: 5.1ha

深圳宝安区石岩水源保护公园景观设计
2008年，中标方案，
用地面积26.7hm²
Landscape Design of
Shiyan Water Resource Protection Park of Bao'an District, Shenzhen
2008 Winning project
Land area: 26.7ha

深圳南澳月亮湾海岸带景观改造概念规划设计
2008年，用地面积14.2hm²
Conceptual Landscape Plannning of
Moon Bay Coast of Nan'ao, Shenzhen
2008 Land area: 14.2ha

深圳市光明新区中央公园景观概念规划设计
2008年，用地面积237hm²
Conceptual Landscape Planning of Central Park
of Guangming New Town, Shenzhen
2008 Land area: 237ha

＊ 长沙楚家湖园景观设计
2007年，用地面积62.5hm²，
景观面积46.88万m²
Landscape Design of Chujia Lake Residential
Community, Changsha
2007 Land area: 62.5ha
Landscape area: 468800m²

＊ 花样年(成都)蒲江大溪谷一期景观设计
2007年，用地面积17hm²，
景观面积12.05万m²，竣工日期：2009年
Landscape Design of Pujiang Grand Valley of
Fantasia Group (Phase 1), Chengdu
2007 Land area: 17ha
Landscape area: 120500m² Constructed in 2009

＊ 深圳三湘国际花园景观设计
2007年，用地面积9.27hm²，
景观面积7万m²，竣工日期：2010年
Landscape Design of
Sanxiang International Garden Residence, Shenzhen
2007 Land area: 9.27ha
Landscape area: 70000m² Constructed in 2010

＊ 深圳龙岗城龙花园四期景观设计
2007年，用地面积3.5hm²，
景观面积3万m²
Landscape Design of Chenglong
Garden (Phase 4), Longgang, Shenzhen
2007 Land area: 3.5ha
Landscape area: 30000m²

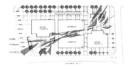

＊ 深圳龙光世纪大厦景观概念设计
2007年，用地面积1.6hm²，
景观面积1.6万m²
Landscape Design of
Longguang Century Tower, Shenzhen
2007 Land area: 1.60ha
Landscape area: 16000m²

＊ 万科宁波金色水岸居住区景观设计
2007年，用地面积13hm²，
景观面积6万m²
Landscape Design of Vanke Golden Waterfront
Residential Community, Ningbo
2007 Land area: 13ha
Landscape area: 60000m²

＊ 深圳金光华春华四季园二期景观设计
2007年，用地面积27hm²，
景观面积8万m²
Landscape Design of Jinguanghua "Chunhua
FourSeasons Garden" (Phase 2), Shenzhen
2007 Land area: 27ha
Landscape area: 80000m²

深圳市罗湖区贝丽中学(水贝珠宝学校)景观设计
2007年，国际竞标，
用地面积1.95hm²
Landscape Design of
Beili Technical School, Shenzhen
2007 International bidding
Land area: 1.95ha

＊ 深圳蛇口君汇新天住宅小区景观设计
2007年，用地面积4.5hm²，
竣工日期：2009年
Landscape Design of Junhuixintian Residential
Community, Sekou, Shenzhen
2007 Land area: 4.5tha
Constructed in 2009

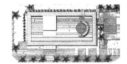

＊ 深圳半岛城邦二期销售中心景观设计
2007年，用地面积2800m²，
景观面积2800m²
Landscape Design of Sales Center of
The Peninsula (Phase 2), Shenzhen
2007 Land area: 2800m²,
Landscape area: 2800m²

佛山南海区狮山镇文化体育公园景观设计
2007年，国际竞标，
用地面积13.9hm²
Landscape Design of Shishan Culture & Sport Park, Nanhai district, Foshan
2007 International bidding
Land area: 13.9ha

珠海前山新冲路城市壹站景观设计
2007年，国际竞标，
用地面积2.85hm²
Landscape Design of the First City Station, Zhuhai
2007 International bidding
Land area: 2.85ha

深圳市龙岗区"深业.坪山"居住区景观设计
2007年，国际竞标，
用地面积2.83hm²
Landscape Design of "Shum Yip Pingshan" Residential Community, Longang district, Shenzhen
2007 International bidding
Land area: 2.83ha

F1摩托艇深圳赛区景观设计
2007年，国际竞标
用地面积70万m²，景观面积70万m²
Landscape Design of Shenzhen Station of F1 Powerboat World Championship
2007 International bidding
Land area: 70ha Landscape area: 700000m²

武汉万科高尔夫城市花园景观设计
2007年，用地面积12.56hm²，
景观面积34.3万m²
Landscape Design of
Vanlke Golf City Garden Residence, Wuhan
2007 Land area: 12.56ha
Landscape area: 343000m²

* 鹏基惠州半山名苑景观设计
2007年，用地面积49.71hm²，
景观面积10.78万m²，竣工日期：2009年
Landscape Design of
Pengji Hillside Residential Community, Huizhou
2007 Land area: 49.71ha
Landscape area: 107800m² Constructed in 2009

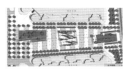
* 深圳南山海德二道海印长城段景观设计
2007年，用地面积1hm²
竣工日期：2008年
Landscape Design of Haide'er Road
(Haiyin Changcheng Part), Shenzhen
2007 Land area: 1ha
Constructed in 2008

* 深圳蓝郡广场景观设计
2007年，用地面积18m²，
竣工日期：2008年
Landscape Design of Lanjun Plaza, Shenzhen
2007 Land area: 18ha
Constructed in 2008

* 深圳沙河世纪假日广场景观设计
2007年，用地面积1.63hm²，
景观面积1.3万m²，竣工日期：2009年
Landscape Design of
Shahe Century & Holiday Plaza, Shenzhen
2007 Land area: 1.63ha
Landscape area: 13000m² Constructed in 2009

* 西安高科城市风景8#府邸景观设计
2007年，用地面积12.56hm²，
景观面积34.3万m²
Landscape Design of
the 8# Mansion of Gaoke City View, Xi'an
2007 Land area: 12.56ha
Landscape area: 343000m²

深圳华侨城"欢乐海岸"总体规划
2006年，用地面积56.46hm²
General Planning of Shenzhen OCT Happy Coast
2006 Land area: 56.46ha

成都市青白江区湿地公园方案设计
2006年，用地面积55hm²
Landscape Design of
Wetland Park of QingBaiJiang District, ChengDu
2006 Land area: 55ha

中央电视台媒体公园景观设计
2006年，国际竞标，
用地面积2.56hm²
Landscape Design of CCTV Media Park
2006 International bidding
Land area: 2.56ha

苏州太湖国家旅游度假区入口节点景观设计
2006年，国际竞标中标方案，
用地面积101400m²
Landscape Design of Entrance Area of
Taihu Lake National Tourist Zone, Suzhou
2006 Winning project of international bidding
Land area: 101400ha

* 成都天府长城2期景观设计
2005年，用地面积5.14hm²，
竣工日期：2007年
Landscape Design of
Tianfu Great Wall Residence (Phase2), Chengdu
2005 Land area: 5.14ha
Constructed in 2007

深圳市蛇口微波山顶景观设计
2005年，用地面积1100m²
Landscape Design of
the Top of Weibo Hill, Shekou, Shenzhen
2005 Land area: 1100m²

* 半岛城邦一期居住区及滨海带景观设计
2005年，用地面积8.36hm²
竣工日期：2007年 获奖项目
Landscape Design of The Peninsula (Phase 1)
and the Seashore Landscape Belt, Shenzhen
2005 Land area: 8.36ha
Constructed in 2007 Awarded project

* 深圳市人民南路环境景观改造方案设计
2004年，用地面积12.44hm²，
竣工日期：2008年 获奖项目
Landscape Design of
Renminnan Boulvard, Shenzhen
2004 Land area: 12.44ha
Constructed in 2008 Awarded project

深圳市沙河世纪山谷住宅小区景观设计
2004年，用地面积18m²
Landscape Design of
Shahe Century Valley Residence, Shenzhen
2004 Land area: 18ha

深圳市鸿荣源龙岗中心城景观设计
2004年，国际竞标，
用地面积40m²
Landscape Design of
Hong Rongyuan Longgang Central City, Shenzhen
2004 International bidding
Land area: 40ha

* 成都华润置地.翡翠城二期及府河公园景观设计
2004年，国际竞标第二名，
用地面积12.70hm²，景观面积2.2万m²
Landscape Design of CRL Jade City (Phase 2) and Fuhe Park, Chengdu
2004 Second prize of international bidding
Land area: 12.7ha Landscape area: 22000m²

* 深圳金光华春华四季园景观设计
2004年，用地面积27hm²，
景观面积8万m²，竣工日期：2007年
Landscape Design of Jinguanghua "Chunhua FourSeasons Garden", Shenzhen
2004 Land area: 27ha
Landscape area: 80000m² Constructed in 2007

西安中海华庭居住小区景观设计
2004年，用地面积5.41hm²
Landscape Design of Zhonghai Huating Residential Community, Xi'an
2004 Land area: 5.41ha

* 深圳市南山区商业文化中心区环境设计
2004年，国际竞标中标方案，
用地面积30hm²，竣工日期：2009年
Landscape Design of Commercial & Cultural Center of Nanshan, Shenzhen
2004 Winning project of international bidding
Land area: 30ha Constructed in 2009

福州市登云山庄景观设计
2004年，用地面积267.4hm²
Landscape Design of Dengyun Mountain Village, Fuzhou
2004 Land area: 267.4ha

湖南天健长沙芙蓉中路项目居住区景观设计
2004年，用地面积17.12hm²
Landscape Design of "Tianjian Changsha Furongzhong Road", Hunan
2004 Land area: 17.12ha

深圳盐田中心区城市设计和中轴线景观设计
2003年，城市规划设计面积18hm²，
环境景观设计面积6.5hm²
Landscape Design of the Center and Central Axis of Yantian District, Shenzhen
2003 Urban planning area: 18ha
Landscape design area: 6.5ha

* 深圳大剧院环境景观改造设计
2003年，景观面积3hm²，
竣工日期：2006年
Landscape Design of Shenzhen Grand Theatre
2003 Landscape area: 3ha
Constructed in 2006

深圳蛇口东填海区大型住宅区景观设计
2003年，用地面积325hm²
Landscape Design of the Large Scale Residential Community of East Reclamation Area of Shekou, Shenzhen
2003 Land area: 325ha

海南省三亚阳光海岸景观设计
2003年，国际竞标，
设计面积146.67hm²
Landscape Design of Sanya Sunny-Coast, Hainan
2003 International bidding
Design area: 146.67ha

深圳湾滨海带景观设计
2003年，国际竞标第二名，
设计范围长15公里
Landscape Design of Sea Belt of Shenzhen Bay
2003 Second prize of international bidding
Landscape design scope: 15km

珠海情侣路滨海带景观设计
2003年，国际竞标，海岸线6.5公里
规划环境景观面积150hm²
Landscape Design of lovers road of Zhuhai
2003 International bidding Coastline: 6.5km
Planning & landscape design area: 150ha

深圳宝安国际机场入口景观设计
2003年，国际竞标，
用地面积30hm²
Landscape Design of Bao'an International Airport Entrances, Shenzhen
2003 International bidding
Land area: 30ha

* 重庆融侨半岛一期2#地块B3区景观设计
2003年，景观设计面积1.75hm²，
竣工日期：2004年
Landscape Design of B3 Plot of Block 2 of Rongqiao Peninsula (Phase 1), Chongqing
2003 Landscape area: 1.75ha
Constructed in 2004

苏州新加坡工业园区东湖大郡二期住宅小区景观设计
2003年，用地面积10hm²
Landscape Design of East Lake Residence (Phase 2) of Singapore Industry Park, Suzhou
2003 Land area: 10ha

深圳市中心区22，23-1街坊街道设施和公园景观设计
2003年，用地面积17hm²
Landscape Design of Streetscape and Park of Block 22,23-1of Central Area, Shenzhen
2003 Land area: 17ha

* 广州玖玖文化家园景观设计
2003年，用地面积4.37hm²，
景观面积3.5万m²，竣工日期：2004年
Landscape Design of Jiujiu Culture Home, Guangzhou
2003 Land area:4.37ha
Landscape area: 35000m² Constructed in 2004

深圳市华侨城玫瑰广场景观设计
2003年，用地面积1.6hm²
Landscape Design of OCT Rose Plaza, Shenzhen
2003 Land area: 1.6ha

中央音乐学院珠海分校景观设计
2003年，国际竞标，用地面积43hm²
Landscape Design of National Music University-Zhuhai Branch
2003 International bidding Land area: 43ha

苏州工业园区东城郡景观设计
2003年，用地面积10hm²，
竣工日期：2006年
Landscape Design of East County Residential Community of Suzhou Industry Park
2003 Land area: 10ha
Constructed in 2006

重庆融侨半岛一期2#地块B1、B2区景观设计
2002年，用地面积11.7hm²，
竣工日期：2004年
Landscape Design of B1& B2 Plots of Block 2 of Rongqiao Peninsula (Phase 1), Chongqing
2002 Land area: 11.7ha
Constructed in 2004

* 深圳市龙岗下沙金沙湾休闲区景观设计
2002年，用地面积17hm²，
竣工日期：2003年
Landscape Design of Golden-Beach Leisure Area of Xiasha, Shenzhen
2002 Land area: 17ha
Constructed in 2003

上海嘉定高尔夫社区景观设计
2002年，用地面积667m²
Landscape Design of Jiading Golf Community, Shanghai
2002 Land area: 667ha

* 深圳中海阳光棕榈园二期景观设计
2002年，景观面积3.3hm²，
Landscape Design of Zhonghai Sunny Palm Garden (Phase 2), Shenzhen
2002 Landscape area: 3.30ha
Constructed in 2003

* 广州瑞丰中大花园环境景观设计
2002年，用地面积33.69hm²，
景观面积1.1万平方米，竣工日期：2003年
Landscape Design of Ruifeng Zhongda Garden, Guangzhou
2002 Land area: 33.69ha
Landscape area: 11000 m² Constructed in 2003

* 深圳市华侨城学校景观设计
2002年，设计面积2hm²，
竣工日期：2003年
Landscape Design of OCT School, Shenzhen
2002 Design area: 2ha
Constructed in 2003

桂林市临桂县九里香堤别墅环境景观设计
2002年，景观面积12hm²
Landscape Design of Jiuli Frangarant Mound Villa of Lingui, Guilin
2002 Landscape area: 12ha

深圳松泉山庄三期环境景观设计
2002年，用地面积1.92hm²，
环境景观用地面积1.6万m²
Landscape Design of Pine Tree and Spring Mountain Village (Phase 3), Shenzhen
2002 Land area: 1.92ha
Landscape area: 16000m²

* 福州融侨锦江A~二期环境景观设计
2002年，景观面积1hm²，
竣工日期：2004年
Landscape Design of Rongqiao Jinjiang A - Phase 2, Fuzhou
2002 Landscape area: 1ha
Constructed in 2004

济南市南部新城区城市中心景观设计
2001年，规划面积396hm²
Landscape Design of the Center of New South District, Jinan
2001 Planning area: 396ha

杭州市钱塘江两岸新市中心区景观设计
2001年，国际竞标，
用地面积300m²
Landscape Design of New City Centers along Qiantang River, Hangzhou
2001 International bidding
Land area: 300ha

深圳星河地产"叠翠九重天"住宅景观设计
2001年，用地面积59.71hm²
Landscape Design of Xinghe Real Estate "Rich Green Paradise" Residential Community, Shenzhen
2001 Land area: 59.71ha

成都长城地产五洲花园住宅区景观设计
2001年，国际竞标，
用地面积63hm²
Landscape Design of Changcheng Five Continents Residence Garden, Chengdu
2001 International bidding
Land area: 63ha

吉林市松花江沿岸总体规划构思及重点地区城市景观设计
2001年，国际竞标，规划陆地面积2525hm²，
城市设计范围971hm²
General Planinng of Songhua River shores and Landscape Design for the Key Areas, Jilin
2001 International bidding Land area: 2525ha
Urban Design area: 971ha

深圳市智泉苑景观设计
2001年，设计面积0.7hm²
Landscape Design of Zhiquanyuan Residence, Shenzhen
2001 Design area: 0.7ha

南海市海八路以北景观设计
2001年，用地面积174hm²
Landscape Design for the North of Haiba Road, Nanhai
2001 Land area: 174ha

* 重庆华立天地豪园景观设计
2001年，环境景观用地面积2.5hm²，
竣工日期：2003年
Landscape Design of Huali World Luxuriant Garden, Chongqing
2001 Landscape area: 2.5ha
Constructed in 2003

* 广州新世界地产凯旋新世界景观设计
2001年，景观面积10hm²，
竣工日期：2004年
Landscape Design of Triumph New-World Garden, Guangzhou
2001 Landscape area: 10ha
Constructed in 2004

深圳龙岗区体育文化广场景观设计
2001年，设计面积2.5hm²
Landscape Design of Longgang Gym and Culture Square, Shenzhen
2001 Design area: 2.5ha

深圳高新技术开发区中心区景观设计
2001年，用地面积25.6hm²
Landscape Design for the Central Part of Shenzhen Hi-tech Development Zone
2001 Land area: 25.6ha

深圳国际网球中心俱乐部小区景观设计
2001年，用地面积9.8hm²
Landscape Design of International
Tennis Center Club Residence, Shenzhen
2001 Land area: 9.8ha

* 深圳中海深圳湾畔花园景观设计
2001年，用地面积2.5hm²,
竣工日期：2002年
Landscape Design of
Zhonghai Seaview Garden, Shenzhen
2001 Land area: 2.5ha
Constructed in 2002

湖州市仁皇山新区城市景观设计
2001年，国际竞标，
用地面积400hm²
Landscape Design of
Renhuangshan New District, Huzhou
2001 International bidding
Land area: 400ha

* 福州融侨水乡温泉别墅景观设计,
2001年，环境景观用地面积11hm²,
竣工日期：2003年
Landscape Design of
Hot Spring Waterfront Villas of Rongqiao, Fuzhou
2001 Landscape area: 11ha
Constructed in 2003

温州苍南新市中心区景观设计
2001年，用地面积454hm²
Landscape Design of
New City Center of Cangnan, Wenzhou
2001 Land area: 454ha

* 深圳深南中路环境及灯光设计
2000年，5公里长，用地面积90hm²,
竣工日期：2000年
Landscape and Lighting Design of
Shennanzhong Road, Shenzhen
2000 Length: 5km Land area: 90ha
Constructed in 2000

深圳市中心公园红荔路以北景观设计
2000年，设计面积36hm²
Landscape Design of Shenzhen Central Park
2000, Design area: 36ha

* 安徽合肥国际会展中心景观设计
2000年，国际竞标中标方案,
用地面积8.30hm²，竣工日期：2002年
Landscape Design of Hefei International
Conference & Exhibition Center, Anhui
2000 Winning project of international bidding
Land area: 8.30ha Constructed in 2002

深圳宝安中心城环境设计
2000年，国际竞标,
用地面积100hm²，绿地面积30万m²
Landscpe Design of
Bao'an Central City, Shenzhen
2000 International bidding
Land area: 100ha Green area: 300000m²

* 深圳市宝安区湖景居环境设计
2000年，用地面积0.7hm²,
竣工日期：2000年
Landscpe Design of Lake View Residence of
Bao'an District, Shenzhen
2000 Land area: 0.7ha Constructed in 2000

* 深圳市城市入口公园景观设计
2000年，用地面积3.5hm²
Landscape Design of
City Entrance Park, Shenzhen
2000 Land area: 3.5ha

* 深圳市盐田区沙头角综合体育中心景观设计
2000年，用地面积0.46hm²,
竣工日期：2004年
Landscape Design of
Yantian Shatoujiao Sports Center, Shenzhen
2000 Land area: 0.46ha Constructed in 2004

深圳莲花山广场环境景观概念设计
2000年，设计面积1.5hm²
Conceptual Landscape Design of
Lianhuashan Square, Shenzhen
2000 Design area: 1.5ha

山东省青岛世纪广场景观设计
1999年，设计面积3hm²
Landscape Design of
Shandong Qingdao Century Square
1999 Land area: 3ha

* 深圳银湖度假区棕榈泉环境景观设计
1999年，用地面积1.50hm²,
竣工日期：2000年
Landscape Design of Palm Spa Area of
Silver Lake Holiday Zone, Shenzhen
1999 Land area: 1.50ha Constructed in 2000

* 深圳规划国地局盐田分局办公楼景观设计
1999年，国际竞标中标方案,
用地面积0.6hm²，竣工日期：2001年
Landscape Design of Yantian Office Building of
Shenzhen Urban Planning & Land Bureau
1999 Winning project of international bidding
Land area: 0.6ha Constructed in 2001

* 深圳时钟广场景观设计
1999年，市政府委托,
样板工程，竣工日期：2002年
Landscape Design of Clock-plaza, Shenzhen
1999, Entrusted by Shenzhen Government,
Model project Constructed in 2002

广西桂林市水系景观设计
1999年，国际竞标,
核心范围300hm²，影响范围600hm²
Landscape Design of Guilin Watersystem, Guangxi
1999 International bidding
Core area 300ha Influence area 600h

* 深圳华侨城OCT生态广场景观设计
1998年，用地面积5hm²，竣工日期：1999年
Landscape Design of
OCT Ecology Plaza, Shenzhen
1998 Land area: 5ha Constructed in 1999

深圳市华侨城中西部城市综合区城市景观设计
1996年，用地面积85hm²
Landscape Design of Complex Area of
Central and West Area of OCT, Shenzhen
1996 Land area: 85ha

特别致谢
Special Acknowledgment

索引

220　2010年客户名录

222　2010年员工名录

Index

220 **2010 Client List**
222 **2010 Staff List**

2010年客户名录
2010 Client List （按首字母索引）

长沙沙河水利投资置业有限公司
Changsha Shahe Water Resource Investment & Real Estate Co., Ltd

成都神州航天房地产有限公司
Chengdu Aerospace Real Estate Co., Ltd

创维半导体（深圳）有限公司
Skyworth Semiconductor (shenzhen) Co., Ltd

鼎和财产保险股份有限公司
Dinhe Property Insurance Co., Ltd.

东海航空有限公司
Shenzhen Donghai Airlines Co., Ltd.

佛山市规划局
Planning Bureau of Foshan

佛山市铁路投资建设集团有限公司
Foshan Railway Investment Construction Group Co, Ltd.

福州市城乡建设发展总公司
Fuzhou Urban and Rural Construction and Development Corporation

高雄市政府工务局新建工程处
Costruction Office, Public Works Bureau, Kaohsiung City Gouvernment, Taiwan, R.O.C

贵阳市城乡规划局
Guiyang Urban Planning Bureau

贵州灵达金跃房地产开发有限公司
Guizhou Lingda Jinyue Real Estate Development Co., Ltd

惠州市仲恺鹏基投资有限公司
Huizhou Zhongkai Pengji Investment Co., Ltd

康佳集团股份有限公司
Konka Group Co., Ltd

宁波万科房地产开发有限公司
Vanke (Ningbo) Real Estate Development Co., Ltd

深业集团（深圳）有限公司
Shum Yip Holdings (Shenzhen) Co., Ltd.

深业南方地产（集团）有限公司
Shun Yip Southern Land (Holdings) Co., Ltd

深业鹏基（集团）有限公司
Shum Yip Pengji (Holdings) Co., Ltd.

深圳和记黄埔观澜地产有限公司
Hutchison Whampoa (Shenzhen) Properties Co., Ltd.

深圳机场（集团）有限公司
Shenzhen Airport(Group)Co., Ltd

深圳机场综合开发公司
Shenzhen Airport Development Co., Ltd.

深圳南海益田置业有限公司
Shenzhen Nanhai Yitian Real Estate Co.,Ltd.

深圳市宝安区石岩街道办事处
Shiyan Substrict Office of Shenzhen Bao'an

深圳市长城物流有限公司
Shenzhen Changcheng Logistic Co.,Ltd

深圳市城市建设开发（集团）公司
Shenzhen Expander (Group) Co., Ltd

深圳市地铁集团有限公司
Shenzhen Metro Co., Ltd.

深圳市规划和国土资源委员会
Shenzhen Urban Planning and Land Resources Committee

深圳市建筑工务署
Construction Works Bureau of Shenzhen

深圳市教育局
Education Bureau of Shenzhen

深圳市科技工贸和信息化委员会
Science, Industry, Trade and Information Technology Commission of Shenzhen Municipality

深圳市龙岗天安数码有限公司
Shenzhen Tian'an Cyber Park (Longgang) Co., Ltd

深圳市罗湖区建设局
Costruction Bureau of Luohu District of Shenzhen

深圳市南山区建设局
Costruction Bureau of Nanshan District of Shenzhen

深圳市南山区建筑工务局
Engineering Affairs Bureau of Nan Shan District, Shenzhen

深圳市青少年活动中心
Shenzhen Adolescent Activities Center

深圳市三新房地产开发有限公司
Shenzhen Sanxin Real Estate Development Co., Ltd

深圳市新天时代投资有限公司
Shenzen Xintian Times Investment Co., Ltd

深圳市玉建房地产开发有限公司
Shenzhen Yujian Real Estate Development Co., Ltd.

深圳市住房和建设局
Bureau of Housing & Construction of Shenzhen

沈阳五里河建设发展有限公司
Shenyang Wulihe Construction & Development Co.,Ltd.

武汉新区建设开发投资有限公司
Wuhan New Area Construction & Development Investment Co., Ltd

香港特别行政区政府
Government of the Hong Kong Special Administrative Region

杨富实业（深圳）有限公司
Yan Full Industrial (Shenzhen) Co., Ltd.

云南堃驰房地产有限公司
Yunnan Kunchi Real Estate Co.,Ltd

中山市金源房地产开发有限公司
Zhongshan Jinyuan Real Estate Development Co., Ltd.

中天城投集团城市建设有限公司
Zhongtian Urban Development Group Company Limited

中天城投集团股份有限公司
Zhongtian Urban Development Group Company Limited

中天城投集团贵阳国际会议展览中心有限公司
Guiyang International Conference & Exposition Center Co.Ltd -Zhongtian Urban Development Group

中铁建工集团深圳投资有限公司
China Railway Construction Engineering Group (Shenzhen) Investment Co.,Ltd

2010年员工名录
2010 Staff List

董事长、主持设计师：冯越强　Feng Yves Yueqiang
董事总经理：何 伟　He Wei
董事设计总监：Michel PERISSE / Christophe GAUDIER
设计、技术、运营董事：王兴法　Wang Xingfa
　　　　　　　　　　沙 军　Sha Jun
　　　　　　　　　　张长文　Zhang Changwen
　　　　　　　　　　白宇西　Bai Yuxi
　　　　　　　　　　杨光伟　Yang Guangwei
　　　　　　　　　　康 彬　Kang Bin
　　　　　　　　　　丁 荣　Ding Rong

曹夜晴	Cao Yeqing	黄 剑	Huang Jian
陈 虎	Chen Hu	黄邦耿	Huang Banggeng
陈 凯	Chen Kai	黄 冈	Huang Gang
陈 然	Chen Ran	黄 恒	Huang Heng
陈 鑫	Chen Xin	黄俊杰	Huang Junjie
陈 勇	Chen Yong	黄启峰	Huang Qifeng
陈成发	Chen Chengfa	黄 茜	Huang Qian
陈康苗	Chen Kangmiao	黄晓晴	Huang Xiaoqing
陈铭恒	Chen Mingheng	黄煦原	Huang Xuyuan
陈双宏	Chen Shuanghong	黄云涛	Huang Yuntao
陈万韬	Chen Wantao	冀 南	Ji Nan
陈小哲	Chen Xiaozhe	简恩洁	Jian Enjie
陈志兵	Chen Zhibing	江志成	Jiang Zhicheng
陈致威	Chen Zhiwei	姜 静	Jiang Jing
崔圣美	Cui Shengmei	蒋伟军	Jiang Weijun
丁生才	Ding Shengcai	焦其新	Jiao Qixin
董孜孜	Dong Zizi	解健民	Xie Jianmin
段安安	Duan An'an	金运丰	Jin Yunfeng
范好仕	Fan Haoshi	景守军	Jing Shoujun
方 翀	Fang Chong	邝英武	Kuang Yingwu
冯 明	Feng Ming	雷 霆	Lei Ting
傅进毅	Fu Jinyi	雷 常	Lei Chang
高笑君	Gao Xiaojun	李 宓	Li Mi
顾 康	Gu Kang	李 瑞	Li Rui
郭 浩	Guo Hao	李翠华	Li Cuihua
郭怀坤	Guo Huaikun	李红芳	Li Hongfang
韩秀兰	Han Xiulan	李开伦	Li Kailun
何 健	He Jian	李绍飞	Li Shaofei
何 明	He Ming	李燕明	Li Yanming
何 伟	He Wei	李 勇	Li Yong
何东婉	He Dongwan	梁 旭	Liang Xu
何红令	He Hongling	林 飞	Lin Fei
何荣山	He Rongshan	林 富	Lin Fu
何同蕾	He Tonglei	林超伟	Lin Chaowei
侯 全	Hou Quan	林法伟	Lin Fawei
黄 刚	Huang Gang	林家统	Lin Jiatong
黄 敏	Huang Min	凌立信	Ling Lixin
黄 媛	Huang Yuan	刘 婧	Liu Jing

刘 威 Liu Wei	王 石 Wang Shi	杨 玲 Yang Ling	Ecaterina Dobrescu
刘长寿 Liu Chang shou	王 斌 Wang Bin	杨 超 Yang Chao	Elsa Burguiere
刘国臣 Liu Guocheng	王丹青 Wang Danqing	杨琳琳 Yang Linlin	Etienne Mares
刘敏婕 Liu Minjie	王和文 Wang Hewen	杨世海 Yang Shihai	Franck CONSTANS
刘琴霞 Liu Qinxia	王瑞芬 Wang Ruifen	杨廷帼 Yang Tingguo	Marie-Christine Freire
刘一锟 Liu Yikun	王莎莎 Wang Shasha	雍兴燕 Yong Xingyan	Marina ZHUKOVA
刘远霞 Liu Qinxia	王锡亮 Wang Xiliang	袁 冲 Yuan Chong	May-Choua VANG
刘志坚 Liu Zhijian	王兴法 Wang Xingfa	岳连生 Yue Liansheng	Valentine Molet
卢东晴 Lu Dongqing	王一军 Wang Yijun	张 恒 Zhang Heng	
陆 兵 Lu Bin	王泽泉 Wang Zequan	张 锟 Zhang Kun	
罗 洁 Luo Jie	王谆民 Wang Chunmin	张昌蓉 Zhang Changrong	
罗长江 Luo Changjiang	王子鹏 Wang Zipeng	张翰宇 Zhang Hanyu	
罗小萍 Luo Xiaoping	魏 婷 Wei Ting	张厚琲 Zhang Houbei	
毛飞霞 Mao Feixia	魏大俊 Wei Dajun	张明珠 Zhang Mingzhu	
毛同祥 Mao Tongxiang	邹 彦 Wu Yan	张英良 Zhang Yingliang	
米歇尔 Michel PERISSE	吴 英 Wu Ying	张峥嵘 Zhang Zhengrong	
聂 凡 Nie Fan	吴纪元 Wu Jiyuan	张志辉 Zhang Zhihui	
欧阳霞 Ouyang Xia	吴俊杰 Wu Junjie	仇福军 Zhang Fujun	
潘智坚 Pan Zhijian	吴 颂 Wu Song	赵 峰 Zhao Feng	
漆文亮 Qi Wenliang	夏 淼 Xia Miao	赵 婷 Zhao Ting	
齐 丽 Qi Li	向 凯 Xiang Kai	赵灵灵 Zhao Lingling	
钱 程 Qian Cheng	肖 浪 Xiao Lang	郑明华 Zheng Minghua	
乔庆丰 Qiao Qingfeng	肖 雪 Xiao Rui	钟洁玲 Zhong Jieling	
丘 蕊 Qiu Rui	肖 琳 Xiao Lin	周 洁 Zhou Jie	
阮慈惠 Ruan Cihui	肖 锐 Xiao Rui	周 茜 Zhou Qian	
沈懋文 Shen Maowen	谢 军 Xie Jun	朱理国 Zhu Liguo	
盛洁菲 Sheng Jiefei	邢彦军 Xing Yanjun	朱婉霞 Zhu Wanxia	
石丹青 Shi Danqing	熊 亮 Xiong Liang	祝 捷 Zhu Jie	
宋 俊 Song Jun	徐 宁 Xu Ning	庄志川 Zhuang Zhichuan	
苏 婷 Su Ting	徐文娟 Xu Wenjuan	邹国强 Zou GuoQiang	
苏 义 Su Yi	许 波 Xu Bo	邹丽施 Zou Lishi	
苏光军 Su Guangjun	许岸程 Xu Ancheng	邹文斌 Zou Wenbin	
孙 晨 Sun Chen	许金华 Xu Jinhua		
孙婷婷 Sun Tingting	严 浩 Yan Hao		
田 静 Tian Jing	杨 安 Yang An		
涂 靖 Tu Jing	杨 力 Yang Li		
王 凯 Wang Kai			

实习生

邓丽山	Deng Lishan
封　颖	Feng Ying
冯　聪	Feng Cong
何思川	He Sichuan
黄国驹	Huang Guoju
季路尧	Ji Luyao
黎裕琪	Li Yuqi
李金波	Li Jinbo
刘承锦	Liu Chengjin
刘子锟	Liu Zikun
罗　辉	Luo Hui
潘占东	Pan Zhandong
钱志慧	Qian Zhihui
商顺明	Shang Shunming
石　灿	Shi Can
孙潺潺	Sun Chanchan
孙宏阳	Sun Hongyang
孙美凤	Sun Meifeng
杨诗冬	Yang Shidong
曾祥林	Zeng Xianglin
张　璐	Zhang Lulu
张一帆	Zhang Yifan
张轶伟	Zhang Yiwei
郑善善	Zheng Shanshan
钟　源	Zhong Yuan
朱　吉	Zhu ji

特别感谢给予Aube(欧博设计)信任的各界朋友、客户，感谢所有全心投入的欧博人。
Grateful thanks to all our friends, colleagues and clients for your trust and surport, as weel as to our devoted Aubers

2010
GREAT EVENTS
大事记

06

2010年6月，<全球视野，走遍世界>欧博考察系列之.2010欧博.世博游
2010年6月，深圳创维"半导体设计中心"建筑方案设计国际竞标第二名

05

2010年5月，成都高新南区综合体项目概念方案设计中标
2010年5月，深圳罗湖档案馆建筑方案设计国际竞标第二名

03

2010年3月，深业东江波尔多皇家庄园概念性规划设计国际竞标中标

2010年3月，贵阳十里花川城市设计与控规（总用地10km², 建筑面积750万m²），获得贵阳规划委员会一致通过

04

2010年4月，佛山西站综合交通枢纽概念性设计国际竞标中标

2010年4月，福州中央商务中心安置房勘察设计方案中标

01

2010年1月，《2010欢乐共享 欧博家家乐翻天》趣味运动会
2010年1月，深圳机场开发区西区即"天空之城"概念性规划设计国际竞标中标
2010年1月，长春南部新城净月西区生态商务金融中心(EBD)城市设计国际竞标第二名（第一名为日本矶崎新事务所）

02

2010年2月，深圳长富金茂大厦项目顺利通过超限高层建筑工程抗震设防专项审查
2010年2月，年度工作会议及欧普利兹颁奖晚会

08

2010年8月，"璀璨夏日清凉游"之惠东、香港湿地公园游活动
2010年8月，深圳君汇新天花园竣工验收
2010年8月，欧博设计与KPF、巴马丹拿合作设计深圳湾一号
　　　　　　（办公、商业、公寓，高度150m)正式启动

09

2010年9月，深圳市高新技术企业联合总部大厦方案设计竞标第二名
2010年9月，贵阳国际会议展览201大厦（生态办公建筑，高度201m)封顶
2010年9月，康佳大厦（总部研发办公建筑，高度125m)封顶
2010年9月，公司副总经理沙军参加《时代建筑》设计品牌传播论坛

10

2010年10月，《我__世界》观念图片摄影展
2010年10月，深圳半岛城邦一期规划、建筑、景观设计荣获"2010年全国人居经典建筑规划设计方案竞赛"综合大奖
2010年10月，深圳南山商业文化中心区核心区环境景观设计荣获"2010年全国人居经典建筑规划设计方案竞赛"规划、环境双金奖
2010年10月，深圳龙岗天安数码新城竣工验收
2010年10月，武汉新区四新生态新城"方岛"区域城市设计国际竞标第二名
2010年10月，公司董事长冯越强获邀参加深圳国信证券大厦国际设计竞赛评标

11

2010年11月，南海中学设计竞标中标
2010年11月，在《di 设计新潮》2010年度最新公布的中国设计市场排行榜中，欧博设计喜获境外榜第4名，较2009年上升3位；总榜名次上升10位，居第44名；效率榜排名上升5位，居第22名

12

2010年12月，欧博人的娱乐组织——"欧乐会（AULE）"诞生！"欧乐会（AULE）"为员工提供自我展示的舞台，通过开展丰富多彩的业余活动促进欧博人的交流，并体验"快乐工作、快乐生活"的欧博文化
2010年12月，贵阳国际会议展览中心主体封顶
2010年12月，深圳机场开发区西区即"天空之城"获2010亚洲国际住宅人居环境奖
2010年12月，冯越强参加华侨城天鹅湖项目整体概念设计国际评标会
2010年12月，公司副总经理白宇西参加第九届深港双城双年展学术委员会
2010年度， 欧博有7位员工步入婚姻殿堂，6位员工晋级为人父母

07

2010年7月，董事长冯越强荣获2010年"中国城乡建设百名杰出人物"奖
2010年7月，欧博设计荣获2010年"中国城乡建设突出贡献企业"
2010年7月，我司设计董事康彬出席评审由深圳市地铁集团有限公司主办的
　　　　　"深圳地铁1号线深大站综合体上盖物业建筑设计国际竞赛"评审会

欧博设计

法国欧博建筑与城市规划设计公司
深圳市欧博设计有限公司
深圳市博艺建筑工程设计有限公司

欧博设计网址: www.aube-archi.com
中国深圳 华侨城生态广场C栋二层
电话:(86 755)26930794 26930795
传真:(86 755)26918376
法国巴黎 1,RUE PRIMATICE 75013 PARIS FRANCE
TEL:(33.1) 45709200
FAX:(33.1) 45709819

Chief Editor: FENG Yves Yueqiang
Executive Editor: GAO Xiaojun
Advisory Editors: BAI Yuxi   YANG Guangwei   Sha Jun   HE Wei   Michel PERISSE
Editors: JIANG Jing   HUANG Gang
Translators: BAI Yuxi   HE Tonglei   LIU Minjie

主　　编: 冯越强
执行主编: 高笑君
顾问编委: 白宇西   杨光伟   沙　军   何　伟   Michel Perisse
编　　委: 姜　静   黄　刚
翻　　译: 白宇西   何同蕾   刘敏婕

图书在版编目(CIP)数据

2010欧博设计/ 法国欧博建筑与城市规划设计公司，深圳市欧博设计有限公司，深圳市博艺建筑工程设计有限公司编著.—北京：中国建筑工业出版社，2011.3
 ISBN 978-7-112-13000-9

Ⅰ.①2··· Ⅱ.①法···②深···③深··· Ⅲ.①建筑设计—作品集—中国 Ⅳ.①TU206

中国版本图书馆CIP数据核字（2011）第037020号

责任编辑：常 燕 袁瑞云

2010欧博设计

法国欧博建筑与城市规划设计公司

深圳市欧博设计有限公司

深圳市博艺建筑工程设计有限公司　编著
*
中国建筑工业出版社出版、发行（北京西郊百万庄）
各地新华书店、建筑书店经销
深圳市国际彩印有限公司印刷
*
开本：787×1092毫米 1/12 印张：19⅛ 字数：120千字
2011年3月第一版　　2011年3月第一次印刷
定价：218.00元
<u>ISBN 978-7-112-13000-9</u>
　　(20378)
版权所有　翻印必究
如有印装质量问题，可寄本社退换
　（邮政编码　100037）